RISE OF THE AVATAR

TIP OF THE SPEAR

JASON BERKOPES

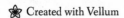

PROLOGUE

Drew Corgain, a biographer and director, grew up wanting to understand how the utopia he grew up in came to be.

What events transpired to bring the world to the event televised in secret to the world?

Who was the man bringing forth the revelation that everyone was being deceived? What was he like? Did he love someone, did he like kicking back with a beer, or did he have a highly religious lifestyle that kept him a pious man?

Those questions kept coming back to him again and again during his adolescence. He found it odd that no person he spoke with ever knew any details leading up to that event.

Drew spent years interviewing Reboot employees, yet information about the man who showed the world it was being grossly deceived was missing.

It felt like the data was intentionally wiped away from the archives, possibly the result of the tumultuous time during which the events took place.

What he did know was that the face of evil was given a name that day.

After finding success with a breakthrough documentary on the worldwide pandemic that killed off most of the world's population, Drew found himself in a position where he could dedicate his time digging up the background on a man few knew much about. Drew spent months convincing an old historian, Robert Bartram, to tell him the story he sought for so long. That opportunity was running out of time.

Robert's health had started declining around a year before Drew met him. After multiple rejections, Robert agreed to the interview after a dream prompted him to change his mind.

Robert would not tell Drew the details of the dream. The fact that he chose to tell the story was incentive enough for Drew not to press for further information.

A small publication studio that Drew owned became the primary location where the audio and video recordings would take place. He understood that time was his enemy, and sending Robert to a studio farther away could end in disaster.

Robert arrived at the studio with an entourage of nurses and assistants to keep him alive. He'd turned down Drew's suggestion that they conduct the interview in Robert's home. The idea of getting out of the house seemed to revitalize the old man to some extent, regardless of the army of medical practitioners it took to do so.

One of the nurses pushed Robert into a moderately sized room with production lights lit around the area. The walls contained ribs with embedded noise canceling circuits used to create the clearest audio recording possible.

Robert sat in his chair that hovered inches from the

floor. A machine attached to the back of the hoverchair contained rows of cylinders flowing different sorts of medications into a centrifuge before being led to a tube going into Robert's chest.

Drew noticed another machine removed what looked like blood and processed it in some way before returning it to Robert. With all the tubes running to and from him, Drew wondered if it might be too late for his interview.

Robert looked at ease in the studio. He found comfort in the familiar hum of electronic devices around the room. More times than he could count, he'd stood in front of news cameras giving depositions on various topics during his life. Though, at his age, he no longer cared about the activities of humanity. He knew his time was at an end, and he was thankful for that.

A recorder hovered some distance away using a technology similar to Robert's chair. Robert's eyes came to rest and locked with Drew's.

"Welcome, Robert. My assistant is going to pin a transmitter to your collar."

Robert nodded with a smile. The assistant clipped the tiny transmitter to Robert to capture and transmit his voice to the camera.

With rapid touches of his fingers to a tablet inset into the arm of his chair, Drew activated the camera and adjusted some settings to his liking.

Robert demanded to be presentable during his sessions with Drew. His midnight blue crushed velvet suit looked as if it were just pulled from the rack at the store. A small white collared shirt and no tie completed the ensemble. To Drew, the look was quite old-fashioned.

"How are you, Robert?"

"Oh, pretty fair considering."

3

"I want to thank you again for doing this. If not for the world ... for my curiosity at any rate. I've wanted to hear this story the moment I saw the intense recording as a child."

Robert nodded. "You are in for one hell of a story, Drew."

Drew looked over the machines keeping Robert alive, remembering how the doctor explained Robert's condition to him.

The machine was used for replacing his blood with blood capable of carrying oxygen when his own would not. It kept Robert alive and allowed him enough comfort to function.

His failing health originated from organ failure at a cellular level, meaning that the cells in his body approached their Heyflick limit. Drew remembered reading about the disability during his college years.

The Heyflick limit is the human cell's maximum number of divisions before the cell cannot divide any further.

The Heyflick scenario was becoming the most common cause of death in Drew's time. Fighting a Heyflick scenario was an unpleasant way to die. Drew had watched his great, great, great-aunt succumb under those conditions. He realized during his communications with Robert that he was dying of old age. There was no better definition.

Robert's body was being ravaged by age, and his mind was losing the ability to remember new things. Yet, his older memories were clear and accurate.

Studies showed many who lay on their deathbed would have sudden improvements in their health.

Those who could no longer hear could hear once more. Those who had poor eyesight could see clearly again.

Drew believed it was God's will to have this story told.

For over one hundred years now, the world has been united under the same banner. Instead of killing in the name of God, humanity created in the name of God.

The common good won out over corruption and greed, but Drew had agonized over the why and how since childhood. The last person in the world who understood the steps it took to get there finally sat in front of him.

As Drew made some final preparations, Robert tinkered with the settings in the arm of his hovering chair, adjusting the height and temperature until he was pleased with them.

Like a seamless dance, the chair's elegant arms swept into the back of the chair as it rose and supported the back of Robert's head.

Drew finished up with his setup and turned to face Robert.

"Robert, I'll have you start with your name, a little bit about your past, and how the world we know of today came to be."

Robert nodded as he gazed up into the lens of the camera. At first, his voice was weak and wispy when he spoke, but it grew stronger as he went.

"My name is Robert Bartram. I was born in a time when war and death were commonplace.

"The concept of war may seem foreign to most of you listening to this. I'm telling this story to assist mankind in understanding what kind of world they originated from. I hope to inspire future generations to never fall into that kind of systematic corruption again.

"Everyone knows the truth. The question I've been asked to answer is why and how. "All of you know the name Reboot. The corporation leads the charge in exploring every type of technology we see today. This wasn't always the case.

"Before Reboot became the world-changing entity it is today, it started like any other small company struggling to be successful in the world at the time. "Back then, companies gobbled each other up for market share like fish in a pond. For a small technology company to survive and grow, they needed to bring something new to the table.

"Reboot's premise was to create a nano-technology that could be injected into the body to fight off disease. The technology took decades to progress to the level it was at when war and pandemics ravaged the world near the middle of the twenty-second century. "If the conflict had erupted a year earlier, humanity might not have survived. If it had, I guarantee this would not be the world everyone enjoys today."

Robert shifted slightly in his chair. A painful grimace appeared for a brief moment before he regained his composure once more.

"I grew up under the tutelage of my father, who was one of the top leaders of the organization, and I followed in his footsteps. After he died in a helicopter crash, I took my place as a board member. "By that point, the company realized another war against power-hungry, corrupted governments was inevitable. Reboot wanted to create a world of peace and science. That dream appeared far-fetched at the time.

"The nanite vaccinations we receive at birth keep us from getting diseases and help us heal faster when injured. "Children are taught about the benefits of nanites in grade school today. In my youth, nanites were on the leading edge of innovation and were made up of theory more than anything physical in nature.

"It takes time to push the technological boundaries of any science, just as it does today. Nanite technology involves a multitude of sciences brought together to make the nanites function. Therefore, the science took decades to engineer—much longer than it takes now with quantum mechanics and other such technologies. "Reboot understood that by taking a stand against the government, they would eventually be labeled as traitors.

"As we considered the evidence, we concluded the United States government was nothing more than a socialist state pretending to be a democracy. "Questionable voting practices were common at the time. Articles about voter fraud were published and then retracted days later. "The government claimed voter ballots could not be doctored, yet they never allowed a third-party entity to audit the electronic voter system to prove otherwise."

Robert paused and took a sip of water from a tube that rested near his mouth and continued.

"Now, the name Reboot was chosen for a couple of reasons by the company's leadership. "First, humanity needed a fresh, incorruptible start in life. In computer terms, humanity needed a reboot.

"Second, rebooting the U.S. government back to its roots was a precursor to giving humanity another chance. All the career politicians needed to be kicked out. New term limitations needed to be established for the judicial and legislative branches, just like we have with the executive branch. We needed to establish the Constitution in such a way as to keep the power with the people. Before our awakening, all governments were corruptible.

"Reboot pulled help from all over the world in achieving these goals. One individual rose out of the crowd to be the tip of the spear against corruption and evil. He was chosen not by just Reboot but by a higher power.

"The rumor of him being blessed by God and obtaining

7

supernatural powers beyond what we at Reboot gave him was confirmed with verified video evidence, which I am providing. "Other rumors of him flying around like a super-hero are false. He would argue that being human is all he ever was."

Drew held up his hand to interrupt.

"You have video evidence of his supernatural ability?"

Robert ejected a small device from the arm of his hovering chair and handed it to Drew's assistant.

"I've stored many videos on this, some of which will blow your mind. You might call me a historian of both the company and the man we all know as Sam. I prefer to just call him my friend.

"The world learned some difficult lessons. Billions died during the pandemic years—billions. Not even the history books of today portray how close we came from toppling over. I want to share how close we came as a human race to extinction, how a corporation you know as Reboot worked hard to save the world, and how a young man named Sam brought everything together."

Robert stopped to take another sip of water. Drew took the moment to ask a question.

"What was the world like before the utopia of today?"

Robert rested his head back against the chair and closed his eyes.

"I need to take you back to the year 2141. Fifty-three years had passed since the end of the last world war. Disease and conflict brought the population from twelve billion to somewhere between one and two billion people based on the best calculations we have."

"What triggered the third world war?" asked Drew.

"The war began as all conflicts did, with someone wanting what someone else had or because of religious persecution. This tug of war by the superpowers kickstarted the gears of war that encompassed the world for the third time.

"From history, we know the French, English, Spanish, Portuguese, Russians, Italians, Mongols, Persians, and many others have all maintained empires at one time or another. "Those nations all fell because they conquered other empires and ruled by fear and subjugation. Until now, no one had ever united the world under a common cause based on peace, religion, and science that all work together in harmony. "When I was in my twenties, saying those three words in the same sentence would have started a brawl. I may have even thrown a punch or two."

Drew chuckled. The old man still had a sense of humor, even at the end.

"Diseases and cancers put constant pressure on the top minds of the time to keep them in check. With pollution rising in the east and the population on a never-ending rise, disease ran rampant. "The World Health Organization and Centers for Disease Control warned the world that a major war might trigger a worldwide pandemic from not just one disease but many.

"Standard infections of the influenza virus proved difficult to contain, let alone the deadlier diseases like SARS, West Nile, and Ebola viruses. "What used to be standard bacterial infections in the 20th and 21st centuries now ate people alive in the 22nd century.

"The life expectancy dropped in every part of the world to levels seen in the early 1900s. The use of vast amounts of disinfectants created immunity disorders in massive amounts of the population. Add in the growing number of diseases resistant to medication, and the tipping point approached.

"It did not take long before the WHO and CDC proved their warnings correct in the years prior to the opening shots in 2088. "The countries at odds sustained themselves for three years before their war machines shut down due to a lack of able-bodied men and women to keep those machines running. Many soldiers died of disease before firing a single shot."

Drew felt himself leaning forward as he listened. "How fast did the death toll rise in the early days of the pandemic?"

"The death toll went from thousands to millions and then to billions within a decade. Huge conveyor belt-fed furnaces burning the dead to ash ran day and night for years around the globe. "Social services broke down in Africa first. Then, the rest of the world followed. With disease centers being overwhelmed with demand or destroyed by war, it looked like an unwinnable situation.

"This was when a small company going by the name Reboot announced they successfully treated one thousand infected patients with a nanite injection containing all the genetic markers of current known diseases. All one thousand patients recovered within days of the single injection. This human study happened in the summer of 2103. "One month later a marketing executive who had taken the vaccination hooked himself up to an IV containing every viral and bacterial disease known to man.

"The news stations monitored him live for thirty days in a quarantine room with no ill effects. Doctors could not find any sign of the diseases injected into his body. "Shortly after, the medical staff announced him as being free of any infectious diseases. Demand for the nanite vaccine exploded, and Reboot became an overnight success."

Drew interrupted for a moment.

"Robert, do you know the executive that chose to be injected with diseases?"

Robert cocked his head and smirked.

"Of course, I do ... the executive was me."

Drew's eyes went wide.

"You volunteered to be injected and quarantined for study like a lab rat?"

Robert nodded. "It was my idea. It wasn't like I could volunteer someone else for the job. In my mind, the technology was rock solid, and I knew the product like the back of my hand."

"Still, that was a hell of a thing you did. Why aren't we taught about things like that in school?"

"To be honest, not many public records exist for Reboot during that time. The government destroyed or locked away most of the company's historical information after declaring them traitorous and all the assets forfeited. "I'll get to the details of how the relationship between the U.S. government and Reboot deteriorated in a moment. Now, where was I?"

"You were just finishing with the part where you were injected with diseases."

"Yes, of course. Prior to that successful test, we ramped up production to give everyone immediate access to the vaccine. "However, tensions between the American government and Reboot rose when Reboot stated they would disperse the vaccination to those in the worst conditions and move up from there. "The powerful and the rich waited in line next to the poor and homeless to get their shots.

"This upset the U.S. government and other world powers. Congress issued an order mandating all dignitaries and government employees be vaccinated prior to anyone else. "But with Reboot's proclamation made public, nothing could be done that wouldn't make them look like an ass, so they capitulated in the end. Many lives were saved because Reboot stood their ground."

"What happened after the vaccine righted the ship?" Drew asked.

"Reboot became a household name across the globe. The next best thing to a cure for disease was Reboot's ability to force the world's governments to work together. "Disease became a distant memory, but the damage caused by the pandemics could not be ignored. Reboot used its new-found leverage to demand certain technologies be made available to improve the quality of life around the world."

Drew held up his hand to pause Robert for a moment.

"I'm not trying to smear Reboot in any way here, but were they always focused on the good, or did they have problems that might shed a darker light on them?"

Robert's chair started beeping. A few of his nurse's leapt into action, opening a panel on the back of the chair, pulling out an empty vial and inserting another full of some clear liquid.

Drew thought it best not to change the subject. He looked over his settings for the audio and video once again to bide some time while the nurses buttoned everything back up and pressed a few buttons to trigger whatever pumps back into action.

Beads of sweat had appeared on Robert's forehead and his face had started twisting into a grimace as he fought off whatever wave of pain the scenario had triggered.

After the alarm was stopped and the new vial inserted into the back of the chair, Robert pressed a couple of buttons and a calm look he had before swept over him.

One of the nurses patted the sweat from his face with a cool cloth and he continued without skipping a beat.

"I am by no means trying to make Reboot out to be perfect. It wasn't. The company used the creation of military technology to help finance the nanite vaccine in secret. "At the beginning of my story, much of society had digressed into how the old west is depicted in historical documents. "People carried projectile weapons out in the open in those days. A similar situation existed in 2141. It was a frightening time to be a law-abiding citizen.

"I'm not trying to dance around the question. I want you to understand that the world today is but a shadow of what it was, including Reboot. "Reboot did what it had to do to accomplish the goals it set out to achieve. That meant doing things that fell into gray areas in order to finance their mission."

Robert paused, causing Drew to look up from his monitor. He thought for a moment before asking, "I'm familiar with the technology between the end of the war and the event the world watched live. "What can you tell me about Reboot from that time? There isn't much public information on how Reboot started."

Robert ran his fingers through his short white hair. Patches could be seen where it had fallen out. His body was dying right in front of Drew. It was hard not to have empathy for the man.

Robert tapped his chin as he thought back.

"Reboot was established over twenty years before the pandemics started ravaging the world. The organization started with three brilliant engineers. Though, I prefer not to mention their names. "The three of them met at the University of Texas on scholarships through the Cockrell School of Engineering. "The university's engineering program garnered much respect in the United States at the time.

"The three engineers happened to all be working on separate PHDs centered on the same subject: nano-technology. "They kept stumbling over one another in the labs, and it didn't take long for them to realize they all had the same focus of study. "They all wanted to make nanotechnology into something more than just a project in a lab.

"Instead of becoming competitive with each other, they became close friends. The three of them collaborated to create the world's smallest injectable robot to assist the body with specific tasks. "The first goal would assist the immune system in defending the body from parasites and disease. If successful, their other goals were to speed up healing and destroy cancer cells in the body.

"The researcher's first goal proved the most difficult. They had to figure out how to get the nanites to identify bacterial or viral infections. "The team also had to find a way to prevent the body's immune system from attacking the nanites. The body would see nanites as invasive and attempt to remove them. The trio realized they required assistance from the bio-medical field.

"The team needed some brilliant biochemists and bio-medical engineers to prevent the body from eliminating the nanites. "Through this realization came the formation of the company called Reboot. At the time, it meant rebooting the human body. Later the name took on a more sinister meaning for the country. "The group used the school's grant money to fund the initial research into the technology. As they all neared the end of their doctorates, they realized they needed to go independent to keep prying eyes away from their research.

"During this time, the world created headline after

headline of near-pandemic outbreaks of SARS, Ebola, Smallpox, and other medication-resistant viruses and bacteria. "This led them to start Reboot as a non-profit research center near their old stomping grounds in Austin, Texas. They had no problems using their contacts in the school system to flush out some research money to pursue cures or other ways to fight the world's diseases. They used the bio-medical research side of the company to cover the underlying research into nanite technology.

"The team covered their tracks by suggesting the research investigate ways to administer future medications more precisely. The cover story never registered as anything suspicious to onlookers. In truth, they gave glimpses into their actual nanite research to the public. They hid the research in plain sight. Nothing seemed out of the ordinary to the casual observer.

"As the years went on, they ran into financial problems because results didn't come fast enough to placate their investors. "However, they achieved two significant break-throughs around this time. The first achievement unlocked the mapping of the bacterial and viral genomes for all the major diseases. The second accomplishment centered around the future personal protection field.

"Once they locked on to the PPF as a side technology, they acquired government funding to develop PPF tech alongside their nanite technology while keeping the nanites a secret. The protective field technology went into prototype with the Navy a few years later."

"Wait ... the government didn't know they were funding the greatest invention in history?" Drew asked.

Robert grinned. "Absolutely not, they would have turned it into a weapon, and the researchers knew it! "We understand that nanites protect us from disease and make us live longer, but nobody outside of the board members knows how the nanites function."

"Won't Reboot object with you leaking that information?"

"I'm not giving you a blueprint for making nanites. I'm just explaining how they function in layman's terms, just as you learned in grade school.

"I don't know half the shit they are pumping in my chest as we speak. I take it on faith that it isn't designed to kill me. That doesn't stop me from wondering how it's keeping me alive."

"I never questioned nanites or wondered how they function at a molecular level. I just know I got a shot at birth and a booster when I turned eighteen," Drew responded.

"You'll get another at fifty and at one hundred. Then you'll likely get to experience your own Hayflick limit like me—unless you experience an accident."

It dawned on Drew he would be one of the last people to speak with Robert.

"I apologize, Robert. I've been harassing you for months, and it just sunk in that you are at the end of your life."

Robert smiled at Drew, then looked down at his hands folded neatly in his lap.

"Boy, don't fret one moment about me. I've lived a hell of a life. Everyone dies at some point, either by accident or design. Mine happens to be by design. I can't think of a better way to go out than this."

Drew sat up in his seat and smiled.

"Then please continue. You were about to explain the process behind the invention of the nanite vaccine."

Robert took a ragged deep breath and continued.

"Let's see—a few months after getting military funding, the researchers discovered a process of powering the nanites by using the thermodynamics of the body instead of pulling energy from the nervous system. "This left the arduous process of imprinting the genetic blueprints of every disease into a microscopic robot.

"Now, I don't know the first thing about how they did any of this. I'm a historian, and I know the internal workings of a functional company. But what they did was extremely difficult. "Imprinting genetic material on the tiniest of robots took as long as the development for the previous processes combined!

"Board members were never allowed access to the details of the technology, even with their high status within the company. Only the handful of researchers who invented the technology had access to that. "The company had to be careful about corporate espionage. People died for much less in those times. "After years of tiny steps forward, the researchers completed successful lab testing on animals.

"The research team submitted their results to the medical field to get approval for human trials. The war broke not long after, and resources became rare. "That also meant the public didn't hear much about the vaccine trials because of the brewing war. To make things even more difficult, one of their labs in California burned to the ground after a nearby nuclear strike.

"Human trials, on the other hand, proved to be an enormous success with no rejections from any human subjects and no side effects. "To prevent rejection from the human host, the nanites received a genetic marker tied to the DNA of the human host they had been injected into. This prevented the body's immune system from seeing the nanites as an infection and allowed the nanites to work in tandem with the immune system. "When the medical review board announced me free of infection, the world went crazy for it.

"As I stated earlier, billions had died from war and disease. Estimates at the time claimed out of the twelve billion people alive near the beginning of the war, approximately one to two billion remained. After some research, I deduced we were closer to one billion than two billion. Between eight and ten percent of the world's population remained. Scary when you think about where we started."

Drew's eyes went wide. "Those numbers are incredible! Are the data points included in the files you gave me?"

"Yes, yes. It's all there. Anything I'm saying is researched and included in the files. I'm not going to tell you anything that I didn't backup with data first."

"When did the government attempt to step in and take control?"

"Well, that wasn't an instantaneous event. Reboot had gone public with the vaccine and with how it was being administered. "It still pissed us off when the government tried to strong-arm us. We had proven that if we administered the vaccine before death, it could save a person, no matter what condition they were in. Those jackasses on the Hill could have cared less. They wanted it first, and we were hell-bent on not giving in to them. I was the public voice of the company by then.

"But the rest of the board was hidden among the company's various positions. Other than the originating five owners, nobody within the company knew who a board member was and wasn't.

"After becoming world-renowned, Reboot shifted from a non-profit organization to a private company. The move prevented the financial records from being made public from that point on. This allowed Reboot to invest in other technology and research with the trillions of dollars it made from the vaccine and PPF designs. All this culminated into the third phase of the company, code-named The Idea."

"What was The Idea? Was it when the coup was developed for seizing power?" Drew asked.

"It didn't start that way, no. At first, the topic in board meetings shifted to the dilemma of what to do about the state of the world and how Reboot could help. "Whole countries ceased to exist. Any remaining countries handled the situations as best they could. "Most countries established different levels of martial law to assist in keeping order.

"People committed murder in the streets. Organized crime took control of many of the abandoned areas of cities. The pandemics left no place untouched.

"Entire countries and even peoples went extinct. It wasn't like Reboot had billions of vaccines ready to administer the moment they announce the vaccine was ready.

"Even if they did, they wouldn't have the distribution or the people to administer the available injections right away. "At the time of my injection, Reboot had maybe one hundred million injections ready to distribute—enough to stop the rise in death toll, but not enough to bring about immediate social services reconstitution.

"The way things were after the war didn't help anything either. The governments that were still functioning gobbled up as much power and land as they could while the nanite vaccine was being administered. That was when the U.S. government overstepped its ground and attempted to force Reboot to hand over the nanite technology. They viewed the microscopic robots as a weapon they could use to hold the world under their yoke, just as they attempted to do with nuclear weapons generations before."

"Seriously? They tried to weaponize them? After all that had happened?" Drew spat.

"Absolutely! Just think, if a government controlled the one thing that protected humanity from disease and for the most part death, they could control every government and every person in the world just by refusing to provide the technology."

"That's insane!" Drew exclaimed.

"That was reality in those times, my boy. I have a copy of a memo between some congressional members that stated the exact thing. Nothing was too contemptuous for the governments of the world. It was the wild west on a world-wide scale."

"What did the company do?"

Robert grunted a bit as he shifted in his chair. He pressed a couple of buttons on the arm of his chair and settled back down.

"As crazy as it sounded, Reboot expected as much and prepared for the tactic in advance. The company sued the government. "The case went to the U.S. Supreme Court, where Reboot won. As the patent holder, the technology remained theirs, and they proved the nanites presented no threat to national security.

"This left a sour taste in the mouths of both sides. Reboot stopped production of their military-grade protection fields for military use. "The government responded by hiring companies to reverse engineer the tech and make it their own. It wasn't as good, but still effective.

"In response to that, Reboot released PPFs into the civilian market. The government attempted forcing Reboot to stop. Yet again, Reboot went to court and used their new power and financial might to gain the upper hand.

"We grew tired of this stupid game. That was when the

idea to overthrow the U.S. government and institute a "Reboot" back to the original Constitution without the hundreds of addendums undermining the power of the people materialized.

"We wanted to put the power back in the hands of the citizens and make some tweaks to prevent any power shift from happening again. "Our sole purpose: to recon each congressional member, the President and his cabinet, and the supreme justices.

"Every conceivable way to research a person found its way into the organization's portfolio. Reboot used their own satellites to spy on the government. The setup took years. Sorting through the data required an entire black site dedicated to the task. "Everything involving the overthrow of the government came through a building called The Grid. All black operations came from that location until Reboot had to go underground completely.

"During this process, a boy named Samson Elijah Jazeel came to the attention of the board members of the company. Sam's actions in that office video you grew up watching triggered a worldwide revolution that made everything we held dear at the time irrelevant. By then Sam and Reboot had gone their separate ways."

"This is the part I've been waiting for. Where did he come from, how did he get chosen to lead the coup, anything and everything you've got on him."

Robert chuckled. His chair bobbed up and down while compensating for the sudden movement.

"This wasn't some global search we did for the ultimate agent. He came to us by pure chance, or at least that was what we thought. "When Reboot started looking for long-term agents, they didn't retrain a person with an established regimen. The company wanted a fresh slate to work from.

"Reboot wanted a person the government affected in some negative way. This gave the person the drive to push themselves, and at the time, we thought it would guarantee they wouldn't turn on us. So, the company started an adoption agency program. The adoption agency fronted as a legitimate business ... with a twist. "As children came in, the agency tested them. The tests provided the levels of intelligence, strength, agility, memory, personality, and temperance the child demonstrated.

"Any abnormally high scores came to the immediate attention of the company and were reviewed. If the child were determined to be a fit for the agent program, the company would adopt them instead of being released to a normal family. "Employees of the company who signed on with the agency took custody of the child and raised them as their own, with specific requirements. The adoption agency found Sam soon after the adoption agency opened for business."

"How did Reboot find him?"

"A pharmaceutical company murdered his parents for trying to publicly announce a cure for a genetic disease the Reboot vaccine didn't affect. "The corporation produced gene therapy drugs used to slow down the progression of certain genetic diseases. Because a cure would affect the shareholders' bottom line, his parents were eliminated, and the cure was labeled a spoof.

"Sam's adoptive parents worked for Reboot. For privacy reasons, we'll call them Jack and Jane. "Jack worked for Reboot in the PPF division. He worked in adapting the technology to smaller and smaller form factors. "He led the team that focused on shrinking the form factor of the PPF down to a level where humans could wear the device for personal protection. "He invented the mechanism allowing high output batteries used in rail gun technology to power the field and make it portable and light.

"Jane worked in marketing. She won a silver medal in the last Olympics prior to the war in long-distance running. That made her perfect for guiding Sam when it came to training his physical abilities, while Jack provided Sam with any mental stimulation he wanted in whatever field he wanted. "They did not force him to be anything he didn't want to be. Instead, they made available any opportunity he showed interest in.

"Sam was one of a kind. The adoption agency discovered that his memorization skills were perfect, and he showed a genius-level IQ. By the time he reached ten years old, he was taking college-level classes. "When he turned twelve, he found out about his biological parents' murder when Jane let the secret slip that he was adopted. Once he discovered the truth, he became even more focused on school. He still loved his adopted parents, but he felt he could do something to help others from losing theirs.

"This is when Reboot approached him about entering the new agent program.

"He would end up being the youngest to be allowed to enter the program. He agreed to the program on the premise that he would be helping prevent others orphaned like him. "In his later teen years, he found it more difficult to push himself in his studies. Sam ran out of things to learn and started being the one who pushed technology forward. He knew nothing of the primary mission at this point. As far as he knew, this physical training program would lead him to other programs.

"Part of his training consisted of an advanced nanite injection he and other agents in training received. The design of the nanite injection would enhance his physical prowess beyond the high levels he already possessed, but being the youngest to be injected with the new nanites became a double-edged sword."

"Whoa, Reboot genetically enhanced their agents?" Drew asked.

"Don't act so surprised. Reboot was living in a world full of predators. One has to be a predator to survive in a world of them."

"So, he wasn't blessed by God? Your serum just enhanced him?"

Robert waved him off. "No, no, you've got it all wrong. I'll get to that blessing craziness later. We just amplified his natural body to match his mind. The injection made him a world-class athlete. If he had attended a normal high school, he would have obliterated every track record he felt like trying for. "By the time he turned eighteen years old, he could run the one-hundred-meter dash in 8.85 seconds. He ran the mile well under four minutes.

"Don't get confused here. He wasn't a superhero like those stories you hear about that were popular before the war. "The nanites allowed the body to function at a level it would have been able to do on its own had a perfect environment been available for it to do so. "The nanites caused his bones to become stronger. He could jump out of trees or out of second-floor windows without injuring himself.

"His medical exams showed he should have grown to be well over six feet tall. Reboot scientists later discovered the density of his bones actually stunted his growth. Tests showed his bones stopped growing by the time he hit 5'8" in his mid-teenage years. "His vision improved to the point he could read six-point font on a monitor from twenty feet away. Most people could not read font that size from three feet away. The downside to his ability to have hawk-like vision was that his eyes didn't adjust well to daylight. "He wore hoods or sunglasses during most of his waking hours. However, his night vision could compete with most predators who hunted at night.

"The nanite injection also modified the programming on the nanites present in his body from his vaccine injection right after birth. They started assisting the body in repairing damaged tissue. At the time, this was not public knowledge. For him, scrapes healed in minutes. A deep cut would stop bleeding in seconds and be fully healed within a few hours. "This type of technology would have been labeled as abnormal and illegal at that time.

"Reboot still retained a moral code. They taught the code to every agent: protect the innocent, leave no person behind, and trust in each other. The agents lived a lie every day after their inception into the agency—they needed a code to anchor them to."

"Did religion play any part during their training?"

"Oh yes, religion played an important part in their development. Some agents showed difficulty in accepting that part of the training. "For Sam, it proved to be a major problem. In his youth, he required more than just a book to believe in God. He asked an instructor once while training on firearms what they believed. "The instructor told him every person needed to find God in their own way. Nobody could tell someone else to believe in something, and then they just have faith. But the question he really wanted answered was *why* they believed.

"The instructor explained to him that all the evidence he would ever need is locked in science. Before the war, science seemed to be on the verge of proving God existed. After the war, almost nobody cared if He did or did not exist. Sam's answers wouldn't come from any living person.

They would come from science. By the time he graduated from the training program, he had found his answer. He became a believer and found time to discuss the notion with anyone who had questions, just as his instructor had assisted him."

"Was he devout by that time?"

Robert shrugged. "I don't know. He was spiritual and had faith. How strong it was in his youth, I cannot say. I *can* say that events transpired later that made his faith unbreakable."

"You mentioned that his foster parents allowed him to pursue his own interests. What interests were those?"

"By the time he reached the end of his first stage of training at eighteen, he had held a master's degree in mechanical engineering, electrical engineering, and computer science. He held two doctorates: one in electrical engineering and the other in mechanical engineering."

"He held that many degrees by the time he turned eighteen? I guess the rumors of his genius are true," Drew said, taking notes.

"He was much smarter than anyone knew. He could apply those skills learned to push the boundaries into new areas. In fact, the day after his eighteenth birthday, he started working with his father at Reboot's PPF facility. He also started the second stage of his training. During his briefing, he found out about the plan to overthrow the U.S. government."

"How did he respond to that?" Drew asked, raising an eyebrow.

"Reboot presented evidence of the government's corruption, what they attempted to force Reboot to do, and what plans Reboot held in store for them. "He saw video evidence of the things the government did they thought no one else knew of. It reminded him of the loss of his parents. He was as dedicated as anyone from that point on.

"Stage two of his training turned him into a weapon. He practiced various martial arts, and he picked them all up with immense speed, but he gravitated to one specific style. Wing Chun became his passion. He liked how it allowed him to attack and defend at the same time.

"Because of the effect the PPFs created in the world, he trained in the different hand-to-hand weapon styles. Melee combat needed to be every agent's bread and butter, but projectile weapons still held a place in the world.

"To counter PPFs, weapons manufacturers made the weapons messy. Large, slow-moving or concussive-based projectiles became favored. Either type tended to make a big mess of the environment they were used in, so melee training became essential in that sort of world."

"What did Sam do when he started working with his father at Reboot?"

"Sam excelled. He made many performance improvements to the PPF design. We later found out that he had made secret improvements to his personal PPF. "His largest contribution was to the team working on a new molecular sword. The blade looked like a work of art—a glass-based metal made up the core of the blade. The core material was then made of a combination of alloys that created the desired strength and flexibility. It gave the sword a very smooth texture, like a mirror. To top it off, an atomized diamond dust epoxy was sprayed over the blade.

"The metal became almost three times as strong as the strongest titanium blade. Yet, the alloy retained a weight equal to titanium. The metal also returned eighty percent of the kinetic energy transferred into it back to the originator.

This made it a capable blade by itself. Two additional features made the sword even more unique.

"First, a high output battery gave the blade an invisible cutting edge, turning it into a molecular blade. The edge of a molecular sword made anywhere except Japan at that time averaged around twenty to thirty molecules wide. That is where the name 'molecular blade' came from. The finest Japanese blades retained edges between five and ten molecules wide. "Sam refined the technology to make the edge only one to three molecules wide. That by itself made it the sharpest blade known to man. But Sam's team didn't stop there.

"The second achievement embedded a sensor into the hilt that would pick up the atomic signature of whatever the blade contacted. Anything touching the blade's surface would be analyzed. Once sheathed in its scabbard, the scabbard would secrete a nanite coating on the naked blade, and the nanites would go to work, making it stronger. This meant the longer a blade remained in use, the stronger the blade became. The nanites also repaired the blade and kept the physical edge razor sharp. This allowed the weapon to remain dangerous if the power cell failed to generate the molecular barrier around the edge of the blade."

"Truly impressive Robert," Drew replied. "What about Sam the person?"

"Sam became a focal point to everything we understand today. The peace and love we see today is because of the hell he lived through and the sacrifices he made. Just as Jesus died for our sins, Sam came to us for a purpose. It took many years for me to piece together the history of Reboot and Sam."

An alert sounded on Robert's chair that brought the attention of his physician and nurse. They moved in a well-choreographed routine of switching out bags of medication and injecting vials of solutions that Drew wasn't sure he wanted to know more about. In the end, he avoided the question and just watched as Robert did his best to smile back at him while they finished up.

Robert made eye contact with the physician and got the OK to continue. He looked back at Drew and smiled for him to go on.

Drew looked at some notes and found the question he wanted to ask next.

"When did you first meet Sam, or did you just monitor him from afar?"

"Ah, yes. I met Sam in person for the first time at his graduation ceremony for our undercover operations. I, as did many of the board members, followed his progress since his adoption program. I still remember the day."

CHAPTER ONE

T he clear blue sky extended on as if the Pacific Ocean hovered overhead. To Sam, that meant a dark pair of sunglasses separated him from the searing pain of the sun glaring into his sensitive eyes—eyes affected by an experimental nanite injection at the age of twelve that had turned his eyes to a pale baby blue.

On the flip side, his vision was as sharp as an eagle's. He was able to see things in the light and the dark that required others to use electronic enhancements to see.

Sam waited in his crisp uniform near the courtyard that occupied the center of the training facility where the graduation ceremony would take place. The facility's courtyard resembled a five-pointed star, like those on the U.S. flag.

In each of the courtyard's five arms, a row of domed buildings with golden roofs lined the edges of the star formation. Each building was a location where Sam learned a skill that made him into a self-sustaining operative. A man of action, as some of his instructors put it.

In a few minutes, he would be graduating from almost two years of rigorous training. But really, he had been training for it since he was around twelve years old. With it came an injection that not only affected his eyes but his body as well.

GPS placed the facility about an hour and a half north of Houston, Texas. The location rested up against Lake Livingston near Hidden Cove Road.

Since the war and pandemics, nature had reclaimed the area. Large trees now dominated the landscape. American beech, red oak, and red maple had retaken where humans once lived.

Reboot, the company he worked for, purchased the land surrounding the lake, including the Sam Houston National Forest to the south. They used the area for training and research. Sam recalled that much of the state and national forests went unattended following the war since lack of funding made maintaining them impossible. Though, after multiple pandemics, there wasn't a need to safeguard forests from the encroachment of civilization.

The state also allowed Reboot to purchase some of the smaller parks for their own use. The forests were free to grow wild again and had done so for the past five decades.

An instructor waved at Sam until he got his attention. Sam joined him, and he led Sam and a handful of other graduates to a small stage. This would be the first class to graduate from the facility. Sam and four others had been rigorously trained before being selected for activation this year.

Coming in at twenty years of age, Sam was the youngest of the five operatives. March 1, 2141 would be a day he would never forget—and not just because his photographic memory wouldn't allow it. He knew something big was coming. He could feel it, like the changing of seasons, the same way the cold air of winter gave way to the warmer air of spring.

A chorus of birds burst from every tree around the courtyard, bringing his attention back to the group he approached. The birds seemed to approve of the graduates and sang their songs in celebration.

Sam was at the top of his class. His performance would be a mark no one would ever reach. Unbeknownst to everyone, this would be the first and last class to ever come through the facility.

Sam recognized all the graduates accompanying him on the stage. Yet, he was never allowed to speak with any of them during training. More than a few of the original twenty-five recruits were failed because they broke that rule. Isolation was a tremendous part of being a field agent.

Other parts of their training put them in situations to observe if they could keep secrets. They were to let no one know they took part in a training program at all. Sam felt part of that was overrated. Another part of him hoped it was overrated—the very ceremony could wind up being a test.

Four males and one female made up the small group of graduates. Sam noticed the young woman on a few occasions around campus but from a distance. He heard she withstood the high physical standards set by the other men and, in many instances, exceeded them.

Her hairstyle remained short yet colorful. The tips of her hair faded to a dark blue on this occasion, though Sam had seen her sporting different colors in the past. The uniform she wore did nothing to hide her natural beauty.

Sam felt an immediate attraction to her. Something about her seemed different from the typical girls that his mother always seemed to set him up with. He overheard her name mentioned, Alyssa Giovnia. It was beautiful.

A few instructors and people Sam didn't recognize spoke some words about how important it was to have this first graduating class, how important they would be to the company, and blah, blah, blah, he'd heard it all before. He had heard it since he was a child. Sam just wanted to get on with it already.

The last speaker was a barrel-chested man with dark, almost black hair and a matching well-groomed beard.

Sam recognized him as the man who survived the experiment with the nanites in the isolation tank on the news when the pandemic was rampaging across the world. It had been some years since, but his face was unmistakable.

Maybe instead of watching the beautiful woman in his class or the birds in the trees, he should have paid more attention to what was being said at the podium. This guy was obviously important.

The man dismissed everyone from the ceremony, and after all the cheers and hoopla died down, he turned to the group.

"Gentlemen and lady."

He smiled at Alyssa, who grinned and looked around at the others as if to say, "Oh yeah, I'm a badass too." Her eyes locked with Sam's for just an instant.

Her cheeks flushed, which made his heart skip a beat. He decided that staring at–Robert Bartram, his name came to him—would be prudent from this point on.

"I would like to say it is an honor to meet the five of you. There is a lot to do. But, in light of the rigorous training regimen and lack of social interaction, we've deemed it necessary to give all of you a few days to relax and blow off some steam."

A tall, lanky young man standing to Robert's right with long brown hair tied in a ponytail at the nape of his neck blurted out, "I'll drink to that!" He looked around absently and shrugged. "If I only had something to drink!"

The group laughed in unison.

"It appears the team has found the class jester," Robert said. "I'll let you all go find some drinks to toast with." He bowed out and walked off. Just before Robert was out of eyesight, he looked Sam and gave him a smile.

CHAPTER TWO

T o change!" Leo said while raising his beer. "To change!" the rest replied in unison.

Sam pounded down his beer with the rest of the team and looked around the bar they ended up at that night. It didn't make much sense as to why any of them were drinking alcohol. The nanite injections they had gotten years ago prevented them from getting drunk.

Alcohol was listed as a foreign substance, and the nanites removed it from the bloodstream before it could affect the brain. It just seemed like the right thing to do on their first night out.

"Does anyone know where we're being stationed?" Sam asked.

"I don't think anyone knows," Tony replied.

"Me either," Alyssa said.

"Hmm."

Sam nodded as if confirming his suspicions.

"I hope we get to work together on occasion. It would be a shame to meet at graduation and never see each other again," Alyssa said, her tone a bit sad.

Alyssa's eyes met with Sam's for a long moment after she finished her sentence. Sam's face blushed when he realized she was talking about him.

In truth, none of them knew if they would. They entered a dangerous world where information became currency, and currency was not to be given away.

"Well, let's take advantage of this opportunity and throw back a few beers and play some darts," Leo replied, moving around the table to squeeze between Alyssa and Sam.

"Alyssa, did you know that Sam here is top of the class? Or so the rumor goes."

She gave him a smirk. "You don't say?"

"I do say. I even heard that he whooped a martial arts master who came here to train him." Leo grinned.

"That isn't the way it happened. It was just a misunderstanding," Sam responded.

"That you resolved by putting your foot in his ass, maybe?"

"Let him be, Leo," Alyssa cut in. "Besides, you sound drunk. Is that even possible?"

Leo turned and beckoned her closer. Her nose wrinkled as the smell of alcohol hit her face.

"I found if you overwhelm the nanites, you can actually get a fair buzz, as long as you don't mind chugging a whole fifth before starting your night. Downside is that you have to pee every five minutes."

Just then, an odd expression crossed his face. "Yep, my five minutes are up!" Leo turned and slapped Sam on the shoulder and scurried off towards someplace to relieve himself. Sam hoped that was a toilet and not the wall outside.

Sam and Alyssa found themselves alone at the table

while Tony and Zack played a round of darts.

Tony was the oldest. He sported a full beard, cut short. At 5'11", he measured around average height for a male his age. Still significantly taller than Sam, though he weighed less than him.

Zack was the tallest of the bunch. Built like a tank, he weighed in close to three hundred pounds and soared over the team at 6'6".

He seemed like Zack ate non-stop since they'd left the facility. From the neck up, he looked like a regular businessman—parted haircut with a trimmed goatee. Tony and Zack seemed to be in their own world, playing darts for money while the others ate and drank.

Meanwhile, the sexual tension between Alyssa and Sam had grown through the night.

Alyssa showed a quality Sam had never noticed in another woman before. It wasn't just her intelligence, appearance, attitude, or humor. It seemed to be all those combined to make her interesting.

Most of the relationships Sam became involved in ended not long after the first few dates. None of them stimulated him long enough to keep his attention. They seemed too busy trying to impress him instead of getting to know him.

Alyssa, on the other hand, kept up with him and kept him guessing. Now they were alone, just staring into each other's eyes.

Sam could tell she wanted him as bad as he wanted her. He reached out and ran his fingers through her hair. He expected she would be asked to change that soon enough.

He liked the look on her. He wanted nothing more than to get her out of those tight jeans and halter top that showed off her washboard stomach.

Sam started to pull her toward him, but she glanced around and leaned away.

"Follow my lead," she whispered.

She walked over to Tony and Zack playing darts.

"Hey guys, I'm heading out."

"What? Why?" Zack asked.

"I've got an early rise. I want to spend time with my family before my orders come in. I don't know when I may see them again, so tell Leo I said bye."

"The night was just getting started!" belched Tony.

"You guys have a great night, and I'll see you on campus tomorrow."

"Alright, see ya, Alyssa."

They returned to their game of darts.

She turned to Sam. "Be a gentleman and escort me back?"

"It would be my pleasure."

Zack and Tony exchanged glances, and with a smirk, they continued as if they heard nothing.

Alyssa dragged Sam out to the street, and they jumped in a transit bus that dropped them out a few miles from campus. They walked the rest of the way, talking about their childhood, life, training, and the stars.

Alyssa slid her hand into Sam's as they talked. Sam stopped her, and out in the middle of a gravel road in the early morning hours, they kissed.

CHAPTER THREE

O ver the next few days, the group stayed together, relaxing, telling stories, and testing each other's abilities.

Leo had not been far off about Sam's hand-to-hand combat abilities. He far surpassed everyone else. So much so that at one point, they all rushed him at once, and that just resulted in some bruises for them.

Sam never rubbed it in. He always managed to twist it so that they learned from how he kicked their ass instead of bruising their egos.

After Sam, Alyssa showed the best grappling talent out of the quartet that remained. She almost put Sam into a submission when they were playing around outside until he reversed the move on her.

When the group wasn't together, Sam and Alyssa were making off somewhere to be alone.

They were nearly insatiable. Maybe it was that they knew they didn't have much time together, or maybe it was love at first sight. Regardless, a fire had started between them that was obvious to anyone occupying the same room.

By the third day, Sam and Alyssa were lounging on each other in the presence of Tony, Zack, and Leo.

The group created a solid bond in the few days they spent with each other. The other three teased the hell out of Alyssa and Sam for sneaking off on their sexual escapades—as if that would slow them down.

Three days after the graduation ceremony, the group was having a picnic in the courtyard.

Leo watched as Alyssa finished eating, laid back and rested her head on Sam's leg, and watched the birds chase each other in the trees. He smiled at the pair of them.

"What's so special about Sam that made you want to be with him? Sure, he's a genius, athletic, knows multiple languages, expert martial artist, expert swordsman, fairly good looking ... shit, I think I just answered my own question."

All of them burst out laughing for a moment.

Alyssa turned her head to look at Sam.

"He's special, Leo. I get him, and he gets me. It's chemistry, so you wouldn't understand."

She smiled at him, and Sam grinned back at her while he ran his fingers through her hair.

"But he's short, almost as short as you."

"That hasn't stopped him from whipping your ass," she quipped.

"Fair point." Leo looked over at Zack and Tony, who were still stuffing their faces.

"You two haven't said much about them slinking around together."

Tony muttered something unintelligible through a mouthful of food and then snickered as if what he said could be understood.

Zack wiped his hands and took a drink of water.

"Shit, man, I have a couple of girls on the side, but I sure wasn't honest with them about what I do. Mixing business with pleasure is never a good idea." He shrugged his shoulders. "Who am I to judge?"

Alyssa picked up her tablet from nearby, and Leo rolled his eyes.

"Sheesh, woman, when are you going to stop looking at that thing for orders?"

She didn't say anything for a moment and just stared at her tablet. She suddenly sat up.

"Uhh, you guys might want to check your mail. I just got my orders."

"What do they say?" Leo asked.

"It says mind your own fucking business and not to open them until in isolation."

Sam stood up, looked at his communicator tablet, and looked up to find Alyssa staring up at him.

"I guess this is it; time for us to go our separate ways."

The other guys noticed the sudden tension and they all dispersed to find out their assignments. This left Alyssa and Sam standing alone in the courtyard.

Alyssa smiled, hugged Sam, and gave him a firm kiss on the lips.

"I had an amazing time with you."

She paused.

"We both knew this wasn't meant to last."

"I know. Time went so quickly. Feels like we just met," Sam said.

Alyssa laughed. "We did!"

"You know what I mean. We've learned so much about each other over the past few days."

"Like I know you have a birthmark on your butt."

It was Sam's turn to laugh. "Not exactly what I meant,

smart ass!" He grabbed her and kissed her. They hugged for a long time until Alyssa gently pulled away. She kissed him one last time, tossed the left-over items into a bucket they used as a makeshift basket, and walked away. She stopped, turned, and waved after a few steps. Sam blushed.

He didn't want it to end and felt a pain in his gut. Had he fallen in love that fast? He shook it off, watched Alyssa disappear around the corner of a building, and headed off in the direction of his quarters.

Alyssa made it past the corner of the building and stopped. A tear fell down her cheek. She let it drop off her chin before wiping the streak away. She just wanted to run back to him.

It dawned on her that she was a trained agent and that there was a bigger picture going on. The country needed her. She took a deep breath and straightened herself up. Checking her makeup in the tablet through the camera screen, she hastened her steps back to her room to find out where she was being sent for her first assignment. She frowned as a realization came to her. They were probably going to make her dye her hair!

CHAPTER FOUR

The sun crept up close to the horizon. Sam had not
been able to sleep the entire night. He just lay in
bed staring at the ceiling of his small cabin clustered among the other thirty or so cabins used to house the
trainees.

He was in turmoil over his mixed emotions. Sam was
excited about the mission, yet his mind kept drifting to
Alyssa, the way she smelled, her eyes, her laugh, the way
she tasted. Back and forth he went all night until his alarm
chimed on his tablet.

He reached over and turned it off, then sat up. Better to
just focus on the job. At least the meeting was almost here.
His orders directed him to enter a building on the campus
he'd never ventured into before.

The sun was just creeping past the horizon as he
approached the small building, a ten-minute walk from his
cabin. He walked with his hands in his pockets, his thoughts
on where he might be assigned and whether he would be
placed with Alyssa or not.

He hadn't paid any attention to the building previously.

He had assumed it was a lawn maintenance equipment shed or something similar.

It stood in stark contrast to the high-tech geodesic dome buildings that ran the length of the facility. They were the only things that showed above ground. The rest of the facility was kept below ground. Even the thousand-meter rifle range lay below ground.

The domes were access points and office space for those running the facility.

On one occasion, he had to infiltrate one of the domes without an access badge as part of a training mission. This building looked much older and may even have been there before Reboot acquired the land.

He gave the door a yank and then a push—it would not budge. He stood back and looked over the side of the shed, noticing a disguised outline of a palm reader to the left of the door. For a moment, the simple appearance of the building had tricked him into thinking it was just a storage shed.

His training told him something strange awaited him. He pressed his palm to the surface, and it lit up green, causing the door to crack open. He entered and found a workbench against the far wall with greasy tools. A table with parts strewn over it stood in the middle of the room. He deduced that the parts were random, which meant they were there for show.

The smooth concrete floor contained two metal inserts inlaid into it, going from wall to wall. They measured in at about four feet wide and were separated by about the same distance.

Next to the table stood a salt and pepper-haired gentleman in black cargo pants and a gray on white battle fatigue shirt. Sam remembered seeing the popular digitized camouflaged print in training videos. Was it a throwback or just an homage to the war?

A sizeable double barrel pistol extended from a holster on his hip. The pistol looked like a concussion-style weapon he'd trained on, a BD-55. The weapon fired explosive rounds used to hurt or disorient the target. Effective, but it lacked the finesse of a molecular sword.

A double scabbard hung from his other hip, and a machete shaped like an old bowie knife rested within it. The blade measured about two feet in length. A combat knife filled the smaller sheath attached to the primary one. Sam noticed the man's belt was actually a PPF—the distinguishable shape that matched his own.

The gentleman extended his hand towards Sam. "Glad to finally meet you, Sam. My name is Darrel Brogman."

Sam shook his hand. "It is a pleasure, sir."

Sam gave another quick inspection of the room.

"Why are we meeting in this place?"

"I'll be happy to answer as soon as ... ahh, here he is."

Sam turned as Leo entered through the door. Darrel introduced himself to Leo and shook his hand.

Leo glanced around for a moment and frowned. "Is this where we are having our briefing?"

Sam had only known Leo for a little over three days, but he liked the guy. He relied on a sly wit and always wore a look on his face that made him appear like he was getting away with something.

Leo and the rest did not get their nanite injections until they reached fifteen years old, which was why Sam remained the shortest of the young men. Leo towered over him at 6'2".

Darrel looked them both over.

"You've both spent years preparing your mind and body

for this stage. If you'll follow me downstairs, I'll brief you on the upcoming mission you'll be tackling together."

The two glanced at each other while trying to figure out what stairs Darrel was referring to.

Darrel walked over near the entrance door, pulled a locket from around his neck, and inserted it into a small hole in the wall. After a few seconds, they heard a click. The metal flooring furthest from the door started retracting. Beneath the flooring laid a set of stairs headed down. He pulled the locket from the hole.

"Be quick now. Follow me."

Darrel moved his way down the stairs. Leo and Sam fell in behind him as the metal shutter started to slide back in place.

"Spy shit is cool," Leo commented as they got to the bottom of the flight of stairs.

Sam smirked.

A state-of-the-art briefing room about thirty feet by thirty feet stretched out before them. A wall appearing to be made of glass faced the stairs.

Darrel turned to them. "Please, after you."

He motioned to the glass door leading into the room. Sam and Leo both walked in and found two interesting chairs in the middle of the room and what appeared to be a tabletop with no legs laying on the floor.

"Please sit. Be warned—the chairs will conform to your body. It will feel a bit odd when they start to adjust. Fidgeting will take them longer to complete their body scans of you."

Leo looked over at Sam in excitement. "Spy shit is cool!"

Sam chuckled. Sam realized Leo's background must not focus on the field of science as Sam's did.

Daniel changed a couple of settings on his tablet. Suddenly, the chairs lit up and started to pulse. It seemed as if they hovered.

A small holographic display lit up on the arm of each chair, showing the biometrics for the person sitting in them. Pulse, body temperature, and blood pressure were displayed in bright blue letters.

Sam remembered reading that lie detectors could be installed into any manner of furniture, turning them into tools for investigating potential employees or interrogating prisoners. Was this another test?

Darrel walked up to the desk, opened an application on his tablet, and pressed a button on it. The tabletop rose from the floor to just above his waist and hovered in place. Wherever he walked, it followed. He started pressing various buttons on the tabletop.

The glass wall at the back of the room turned a solid black. A holographic display beamed on the wall showing the Reboot logo.

The logo was shaped like a green arrow that curved around in a circle until it pointed back at its starting point. A solid blue sphere stood alone in the center. The sphere represented Earth.

"Give me a moment while I get the briefing material ready."

Darrel pulled some things out of the black metal briefcase sitting on the hovering table. That gave Sam a moment to test his theory on the true function of the chairs.

"Leo, I'm going to ask you a question. I want you to give me a false answer."

"Huh? You want me to lie to you?"

"Yes."

"OK."

"What is your name?" asked Sam.

"My name is Alyssa. Ohhh, and I love my man, Sammy. He is so smart and so strong," Leo replied, batting his eyes.

The chair did nothing other than hum and continue displaying Leo's biological data.

Leo's joke about Alyssa didn't go unnoticed by Sam. He had just gotten her out of his head, and Leo pulled her right back in. He decided to ignore Leo and gave him his best poker face.

"Well done, Sam. You figured out the chair's function. That feature is not active. Would you like to see what it does?" Darrel asked.

"As a matter of fact, I would," Sam said with a wry smile.

Leo whispered to him. "Spy shit is so cool!"

Darrel tapped a couple of buttons on his tabletop that floated in front of him. "OK, ask Leo the same question."

Sam turned back to Leo. "Leo, what is your name?"

"My name is Alyssaaaahhhh hey! This thing shocked me!" Leo leaped out of the chair.

Darrel and Sam almost doubled over in laughter. "Is spy shit still cool?" Darrel asked.

"Well, most of it is. I'm not certain about these chairs anymore."

"These chairs are used to interview potential employees or to interrogate people. They contain integrated lie detectors. The shock is an option during the interrogation process."

Leo raised an eyebrow at the man. "You shock people during interviews?"

"Don't be ridiculous. We don't shock people coming in for interviews. We just shock trainees who aren't focusing."

Sam and Leo went wide-eyed at the last comment.

"OK, enough fun, men. Let's get to the real reason you're here."

Both perked up at that statement. Leo hesitantly sat back in his chair.

"You two are being paired together for an infiltration mission of the highest importance. Leo, you excelled at speech and math. You also have a knack for persuading others to believe in you. You are easy to like and easy to forget. You are being tasked with running for a position in the House of Representatives coming up this year as candidate Jake Tillman. We've taken steps to make certain Mr. Tillman wins the election. The moment you leave this building, the name Leo Ferdinan McKiny will cease to exist. Further details will follow."

Leo smiled. "I've never aspired to become a congressman, but I did want to be an actor growing up. I guess this is my big break!" Leo pointed towards Sam. "So, what is Sammy doing? He running for President?"

Darrel glanced down at some information crawling across the screen of his monitor.

"Sam, your primary mission is to protect the congressman here." Darrel pointed in Leo's direction.

"Whoa, Sam gets to babysit while I do all the heavy lifting? Sounds boring," Leo commented.

Darrel cleared his throat. "Your secondary mission will be to recon any other congressmen that meet with Mr. Tillman. You're also to attend to any other contacts that may come into question."

"Never mind, Sammy's job is harder."

"Understand this—not everyone on the Hill is corrupt. Some are being coerced, and others are too afraid to do what they know is right. Those are the ones we need to protect and have on our side. I suggest you both settle in for a long day. I'll be going over the details and schedule on how this is going to take place. Everything we have on the mission will be sent to your Personal Data Assistants, which I'm going to give you. They are biometric and require finger, retina, and a facial recognition scan to unlock them."

He looked to Sam and Leo. "Any questions before we get started?"

They both shook their heads no.

"Let's proceed."

CHAPTER FIVE

During the next couple of months, Leo morphed into his alter ego, Jake Tillman. He was trained in foreign and domestic politics. He spent countless hours wrapping his mind around the two political parties, their history, what their goals had been, and what they currently are.

He was introduced to his campaign manager and other team members working to get him elected.

His final process was to present himself as a typical politician. From the way he dressed, held himself, and spoke.

Sam spent most of his time training in the art of being a bodyguard and managing a team of security agents to guard Jake on his campaign.

The new Jake Tillman and Samuel Miller met each week for a grappling session to keep their skills sharp and to continue to bond. Sam and Jake stretched out on the grappling mat in a dojo at Jake's home. Reboot had set the place up as his backstory.

A large monitor hung from the corner of the dojo,

showing the news with the audio muted so that Jake could stay current on what was going on even while training. Their half-naked bodies and colorful tight-fitting shorts would have come across as a burlesque show had they not been standing in a gym.

Their chiseled muscles rippled as they stretch in preparation for their session.

"How is the training coming along?" Jake grunted as he grabbed the bottom of his feet, giving his back a good stretch.

"I've been training with the security team that is guarding you. I feel ready."

"Ready? From what I've heard, you come across as a seasoned veteran, like you've been doing security for years."

"I'm sure the guys are just being nice."

Jake laughed. "When have you known the guys to be nice?"

"Good point."

"What do you think of our chances?"

"Chances of getting you elected? I'd say you are a lock now that we have evidence that the congressman you're running against has been committing adultery with a minor."

"No, I mean ... you know, our chances at the primary mission."

"I try not to think about the big picture. With so many moving pieces, it is hard to say. I have faith we'll win out."

Jake rolled his eyes. "Faith? Don't start with that again."

"I know you don't believe, but I know in my heart that if God is willing, we will succeed."

"You know that people will die during this coup on both sides, right? Yet, you think an all-knowing, loving God is going to condone that action?"

"God has done much worse in retribution in the past, Jake. That doesn't mean he lacks love or empathy for us."

"I don't think that holds much water."

"It does hold water; you just refuse to see it."

Jake grabbed a roller and started rolling out his leg muscles. Sam shrugged and stood up, then went over to hang from a pull-up bar mounted to the wall to stretch out his upper back.

"Look at the world, man. This is no Garden of Eden. It is a shit show at best. You and I grew up orphans. Our parents were slaughtered, and we were raised by a company that wants us to overthrow the government. Tell me that isn't fucked up."

"It depends on our point of view. I'm in this to right a wrong, the wrong done to millions of us over generations by a corrupted government."

"You think us replacing the government is going to change things for the better?"

"I do, because I have faith."

"There's that faith thing again. You may want to keep that stuff to yourself. If the ops folks at Reboot hear you talking that nonsense, they will pull you from the mission."

"Our handlers at Reboot know full well my thoughts on religion and God. I've had long discussions on the subject with many people there."

"Maybe that is why you are on babysitting detail and not in my position—ever think of that?"

"I am willing to do whatever I'm tasked with. I don't have any aspirations of power."

Jake shrugged. "I'm here because it keeps me off the streets."

"To each their own, brother."

Jake jumped up and started dancing around, shaking out his arms.

"I'm feeling good today, buddy boy. You ready to go down?" Jake gave Sam a wink and a smile.

Sam dropped off the bar to the ground and sized up Jake. On paper, Jake's stats said he should have dominated Sam, with his longer reach and size advantage.

"I'm feeling pretty spry myself."

Sam winked back at Jake, and they faced off.

Thirty minutes later, they were soaked in sweat and Jake's back was red with how many times he had been on his back. Sam's back was as pasty white as when they started.

They took a break to go towel off and hydrate.

"That was a good workout today. You almost had me a couple of times," said Sam.

"I've learned a lot from you. I'd like to think I'm getting better."

"You definitely are."

Jake grinned. "Thanks."

Jake gave Sam a playful shove.

"Sorry about earlier. I just don't know how you can have faith in someone you can't see or hear."

Sam laughed.

"Oh, I can see and hear Him, Jake."

Jake looked around. "How?"

"I'm a scientist; you think I haven't had a heart to heart with myself over who God is and whether or not He exists? Science has shown me that He does exist. In my mind, it's irrefutable."

"So, no changing your mind, eh?"

"I suspect you're asking me these things because you're curious, maybe wondering if I'm right?"

"Nope, I think you are bat-shit crazy." Jake laughed.

Sam smiled back at him. "I am not going to tell you that you have to believe or even try to convince you. I will tell you what I believe, just as I've done since I've known you. I'm here if you ever want to explore more of that side of the coin."

"Maybe, but for now, I'm more comfortable with you being the delusional one."

"Let's shower up and get ready for that state dinner tonight," Sam suggested, shrugging off the mental comment.

"This election is going to be a cake-walk now," Jake replied.

"I think I'll have a harder time winning over the security team in D.C. than you will at winning this election."

Jake grabbed Sam's arm to stop him. He locked eyes with Sam's.

"About that ... I've heard some things about Capitol Security, not good things. Be careful when we get to D.C. Those guys are animals."

Sam nodded. "I'll find a way to win them over."

"It certainly won't be with your size."

Sam reached on his tip toes.

"If I walk around like this, you think it would help?" he said jokingly.

"Only if you want them to think you're a ballerina!"

They both laughed as they walk through the door to the showers.

CHAPTER SIX

S am sat outside of the building that housed the capital security forces for the city. He took off his glasses and let the daylight force his eyes to water.

It reminded him of the sacrifices that were going on behind the scenes—those of his colleagues around the country.

Reboot's goals did not include controlling the country. If they succeeded in overthrowing the corrupt government body, Reboot planned on instituting the necessary changes to prevent the corruption from happening again.

People were elected all the time that wanted to make positive changes to the government. That wasn't the problem.

One of the tasks for Sam was to determine why those people changed so drastically after being elected.

The corruption of the government came from loopholes found out from within. Sam needed to find out who was controlling the government. How many corrupt members existed, and how did they exert influence upon the other congressional members?

Many politicians owed their elections to powerful lobbyists who paved the way to Capitol Hill with a highway of money.

Reboot wanted to outlaw any and all means of financial donations to campaigns or otherwise. They wanted the government to regulate the amount of money allowed and to fund the campaigns. That would prevent illegal donations and other activities from rewarding congressional members for siding with the lobbyist.

The final reason Reboot wanted to stop the government steamroller was to prevent them from creating an unconstitutional law allowing them to seize all technology from not just Reboot but from any institution they deemed influential.

The process to fight a law or bill thought to be unconstitutional would take months or even years for the lawsuit to be presented to the Supreme Court for a ruling. That would be enough time to retrieve the technology secrets they wanted before the Supreme Court could overturn the bill, and there were no guarantees the bill would be overturned at all.

Reboot wanted to prevent the government from weaponizing their nanite technology at all costs. If the military obtained a weaponized nanite technology, they could threaten the world with a weapon more devastating than a pandemic—it could be programmed to kill.

Everyone taking part in this coup knew the consequences, as did Jake and Sam. Now that Jake held a congressional position, he needed to settle in and learn as much as possible about how the real process worked versus the letter of the law.

Jake found out fast what the priorities of the capital were. His first meeting discussed his appointments with his personal staff. He did his best to play along with those whispering in his ear on how to proceed.

A specific security company provided the security for all congressional members, which should have been an illegal monopoly. Yet, somehow, it was overlooked. When it came time to choose the head of his security detail, he, of course, chose Sam. So, bringing in an unknown to be head of his security detail raised some eyebrows.

Jake smoothed some ruffled feathers by explaining that Sam protected him during his entire election campaign. There was a comfort zone between the two of them Jake enjoyed. That provided enough of an excuse to ease the egos of the upper staff.

That put some burden on Sam to garner trust and respect as fast as possible to get their eyes off him. That way, he could go about collecting data. Jake's statement during their covert briefing of the mission seemed truer now than intended. Sam was an unknown. He wasn't a company man, and Sam didn't fit the typical bodyguard profile.

These capital bodyguards exemplified the word 'immense.' The average weight tipped the scale around 250 pounds, and height reached well over six feet tall. Sam weighed in at a lithe 190 pounds and only reached 5'10" with boots on.

Time to prove my mettle, Sam thought to himself while sitting on the bench under a shade tree. He cleaned himself up and headed inside to get the ball rolling.

Sam called the squad together in an oval staging area lined with lockers and racks of gear along the walls. An armory was located at the far end away from the briefing area lined with rows of tables and chairs and a podium with a grid of the city laid out in miniature hologram where they met to go over the congressman's schedule, travel routes, and security coverage for the coming day. A vomit-colored paint gave the room an air of unpleasantness.

When Sam entered the room, he noticed the negative looks and gestures coming from the group. He tried his best to be professional, yet he garnered no respect from the men who answered to him.

He wore his molecular sword on his back and positioned it at an angle so that the hilt rested near his right ear. A sidearm hung from his left leg that fired 20-gauge explosive slugs, and another hung from his right that fired three-inch armor-piercing rounds designed to penetrate PPFs. He marched through the men lounging around the room and up to the front.

"Apparently, we have a problem to resolve. Yesterday I ordered the team to train breaching tactics, and nobody showed," Sam said.

Brad, a perfect definition of the stereotypical security team member, rose from the desk he sat on. The desk groaned from its joints as the massive man lifted his weight off of it. He puffed his chest out and stared down his nose at Sam.

"What problem are you referring to? The fact we all hate you or that you are a little bitch?"

Chuckles erupted from the group behind him.

Sam smirked. This was the guy he was going to make an example of.

"Regardless of how you feel about me, I am still in charge. If you don't start listening to and obeying my orders, I will replace you."

"Replace us? With who? Nobody in this town wants to work with you. We run all the security in D.C. Just because you know the congressman and sucked his dick to get a cushy job doesn't mean shit to us."

Sam laid down his clipboard on a nearby desk and walked up to Brad. Brad puffed out his chest even more and stood at his full height. He towered over Sam by at least six inches. Sam just stared up at him, steadfast.

"You act as if you're speaking for everyone."

Sam leaned to the side to peer around Brad at the rest of the men in the room.

"Is he speaking for the rest of you? Am I a little bitch that sucked some dick to get a cushy job?"

They all glanced around at each other for a moment.

"Well?"

Tory, a well-built man with a chiseled chin and expressive eyes, cleared his throat. "Sort of, yeah. We don't know you. You seem like you're out of your element and need to go back to wherever you came from. Let us do what we do best."

Sam looked back up at Brad.

"I get it. I do."

He turned and took a few steps back towards his clipboard.

"You see me and think I'm not up to the task and that I don't belong because I don't fit in."

He turned back towards Brad and shrugged. "Maybe a test is in order?"

"Like what?" Brad spat back at him.

Sam thought for a moment.

"I'll give you the first punch. If you drop me, I'll quit, and you can all take over security duties. However, if I drop you within fifteen seconds of you throwing the first punch, you are out, and the rest of you start listening to me. Deal?"

Brad scoffed. "Are you serious?"

"Dead serious."

The entire room burst out in laughter.

"Like, right now?"

"Sure, why not?" Sam replied, looking around at everyone still laughing.

"Shit, this day is turning out excellent."

One of the guys clicked on his wristwatch to start a timer. Brad turned back to Sam.

"Here it comes, little man."

Brad expected Sam to be proficient in some hand-to-hand training, so he readied himself, gave a little feint jab, and followed it with a powerful straight right at Sam's face. Sam slipped to the side and caught Brad's wrist with his right hand while bringing up his left into Brad's throat with a lightning-fast palm strike.

Brad's breath caught in his throat. Before he reacted, Sam, still holding Brad's wrist in a vise-like grip, did a front flip and brought the back of his heel down on Brad's head. The big man came crashing down on the floor. Sam, still holding Brad's wrist, rotated Brad's arm into a lock behind his head.

"Time?" Sam asked casually.

Tory glanced down at his watch. "Holy shit ... you took seven seconds."

Sam let Brad go. Brad rubbed the top of his head and stumbled to his feet. Gary, another team member, jumped up out of his chair. "That's the craziest shit I've ever seen!"

Sam smiled and turned back to Brad. "Brad, get your belongings. The door is back that way, big fella."

Brad growled back at him. "Screw you, man!"

He picked up a nearby chair by the backrest, and with one arm, swung the heavy chair in an arc at Sam.

Sam reacted in an instant, pulling his sword from its sheath and cutting the chair in two in a single downward swipe of the sword. In the next instant, he swung the sword in a quick flash at Brad's waist before sheathing it back into its scabbard.

Everyone jumped up when he did that. Brad froze. He couldn't believe how fast Sam moved and was afraid to see what he had done to him.

One of the guys gave a chuckle. Brad looked down to see a slice right through his belt and pants. His pants started to slide down, and he snatched them back up and glared at Sam.

"As I said, you can gather your things and go. I wouldn't suggest doing anything as stupid as you just did again. If any others are so inclined, they may leave as well."

Tory, who had timed Sam earlier, looked back at his watch and then to Sam again. "Hell no! I'm staying!"

The rest of the team nodded in agreement.

After Brad left, Tory stood in front of the team.

"Listen, we just thought you were some bureaucrat they assigned to congressman Tillman because you were friends, and he was just trying to get you a job. We take this job very seriously and thought you didn't have a skill set that would allow you to handle this position. On behalf of the team, I apologize for our actions."

"Apology accepted. How about us getting over to the training facility and showing me your skills at breaching tactics? I'll show you what I know, and we can collaborate on any ideas about covering the congressman during his movements around the country."

"That sounds great!" Tory exclaimed, and the team grabbed their gear and headed out.

Sam lingered a bit, picked up the debris from the desk, and tossed it in the corner. He smiled to himself and grabbed his gear before following the team out. When Sam met with Jake later, he explained the situation. Jake laughed over the ordeal.

"Your first true test wasn't protecting me. You had to protect yourself first!"

Sam smiled. Then, his face turned sincere. "I want to apologize. I know keeping me here put you under a microscope."

"Not to worry. All part of being a rookie on campus, I suppose." Jake gave Sam a look that signaled to him they were probably not alone.

"I have briefing after briefing for the next week. It should be a light week as far as security is concerned—it sort of feels like college all over again. They cram in so much information, it's crazy. Now I know why I hired all the staff members. No one person could keep this straight! The situation makes me wonder what the country must have been like in the early years."

"We can only hope that being a congressman was less of a headache in those days."

Sam knew the numbers. Crime was out of control because so many of the cities were abandoned. With a fraction of the population that existed before the war, managing a crumbling infrastructure couldn't be done. Yet, the government didn't lift a finger to improve the situation. It seemed that Capitol Hill followed the business-as-usual mantra.

CHAPTER SEVEN

J ake and Sam were as brothers. Months of training had brought them closer than they even expected.

Being in the political eye, they grew accustomed to keeping tabs on what they said. They could switch between speaking in code as easily as they spoke normal English.

They both dealt with a lot of stress and handled it as well as they could, relying on each other often. Jokes, teasing, and everything else two brothers would do, they did. It was similar to how people from different backgrounds bonded in a wartime environment. They didn't fight for the cause so much as for each other.

About two months after his inception, Jake was invited to a meeting. Sam accompanied him up to the entrance of a refined condo in D.C., where he was not allowed to enter.

Sam remained back with the vehicle, took notes, and monitored Jake with a tracker he placed on him before dropping him off. Even Jake didn't know Sam monitored him from the vehicle. Due to the risks, Sam couldn't use any audio —it was too easy to scan for. Instead, he stuck with knowing where Jake's locator showed up in the building. If for some reason, his location moved to an area Sam didn't believe Jake should be, Sam could react and attempt to reach him.

After Jake stepped inside, a large man patted him down while another scanned him. The condo's foyer was three stories tall and had a gaudy chandelier hanging from the ceiling, projecting bright spots along the walls like a disco ball. A piece that expensive did nothing but show off power and money. *Must be nice not to have a conscious*, Jake thought.

Expensive wood trim and flooring led off into other rooms. Velvet upholstered chairs and antique mahogany tables lined the hallway that disappeared into the darkness beyond.

"What is the scan for?" Jake asked.

"Just a precaution to make certain you did not get bugged without your knowledge, Mr. Tillman. Everyone goes through a scan."

As he finished up, he heard a door open behind him. Jake turned to find an older gentleman coming through in a classy business suit. He looked like the stereotypical politician, wearing an expensive blue suit with a white collared shirt many of the congressmen wore. It seemed the one thing differentiating them from one another was their tie.

Clothes couldn't be too colorful, or they would stand out. The suits reminded Jake of a flock of birds. When seen together in a group, it was difficult to point out a specific congressional member from the flock.

The gentleman reached out and shook Jake's hand.

"Hello, congressman Tillman. It's great to meet you. We've been meaning to meet with you sooner. All those meetings they put the freshman in to bring you up to speed make it difficult to find time to pull you away. My apologies; I failed to introduce myself. I'm Senator Lambert, fellow Texan and a big fan of you."

Jake shook his hand. "How so?"

"Your campaign was a major upset. A Democrat had filled the slot for more terms than I wish to think about. It was great to see a Republican snag the seat back to make the state one hundred percent Republican again. But what really impressed us was the style you used to defeat your opponent."

"I just made the promises he couldn't make. And ... well ... it helps when he is caught with a seventeen-year-old prostitute in his limo."

Lambert nodded. "Ahh, yes. The public eye tends to open wide when things like that bubble to the surface. That is something you need to make certain of. Do not, under any circumstances, let the public know about your dirty laundry." The senator winked at him and gave a quick well-practiced smile.

"I'll be certain to remember the tip."

"Let me introduce you to the rest of my colleagues here in the other room."

He led Jake towards the door he'd exited earlier.

The two entered an immense library that shot up as high as the foyer he came in through. In the center, a vast round table rested with many expensive leather-bound chairs lining it. A group of fellow politicians clustered near one side talking and laughing. They sipped brandy and smoked cigars. The bitter scent of cigar smoke stung Jake's nose. The men noticed the two approaching and stopped talking, sat down their brandy and cigars, and stood to face Jake and senator Lambert. Jake surmised the men could be in their 50s to early 60s. They must be the movers and shakers Jake heard about.

"Gentlemen, I would like to introduce Congressman Jake Tillman. Jake, starting from your left is Senator Browling from New York, Senator Mantel from Florida, Senator Tompkins from Illinois, Senator Patterson from California, Senator Christianson from Texas, and last but not least is the speaker of the House, Gregory Crocker."

Jake shook their hand as they were introduced to him.

"Please take a seat. Would you like a cigar or some brandy, Mr. Tillman?" Senator Christianson asked.

"I'm fine, thank you."

Jake settled down into the plush leather chair and crossed his legs. The others followed suit. A couple of them grabbed their cigars and brandy from the table where they set them. Cigar smoke rolled out above them like some old coal-fired locomotive Jake came across pictures of in school.

The men stared at Jake with smirks on their faces. Jake could tell they were up to something. Their eyes darted about like they were cats looking for a mouse. At that point, he knew this was no ordinary meeting. The situation seemed to be about propositioning him with something to test him.

Senator Lambert spoke first. "As you know, the United States, along with most of the world, has been crippled by war and disease. Over the past decade or so, we've relied on technology from a company named Reboot to bring us back from the brink. Are you familiar with the company?"

"Of course ... Everyone knows about Reboot. They invented the nanite vaccine. They also provided our military branches with protection field technology, allowing us to thwart new projectile technologies. The company is the wealthiest in the history of the world. Rumors suggest there became some disagreements between the government and Reboot over some rights to technology. That is the extent of my knowledge of the company."

The senators all stared at him with blank poker faces. Senator Lambert leaned forward.

"Jake, our relationship with Reboot has been a tenuous one. They gave us full rights to their protection field technology for military hardware, at least until our disagreement over how to utilize the nanite tech. They publicly denied our government officials priority to the nanite vaccine when they distributed the initial doses. The situation endangered the welfare of our country."

Jake interrupted. "I thought they prioritized vaccines by the status of the patient not the status of the person in society."

The senator continued unabated. "The point I'm trying to make here, Jake, is we don't have control of our own country. Reboot is holding us prisoner by refusing us the technology we need to keep the citizens safe."

"I think I'm starting to see where this is going," Jake said. "By law, they have the right to deny you technology based on patents and other privacy infringement laws created to protect inventors of technology."

"For now ..." he replied. "We put into motion a bill we plan on passing that will allow us to acquire the technology from them. That is where you come in."

"Me?" Jake replied as he cocked his head. "What is a freshman congressman going to be able to do that you can't accomplish on your own?"

"To be blunt, we need your vote, and we need it soon. Reboot is fighting us by pumping money into lobbyists and turning the ears of other senators and representatives like you against us."

"I see. Isn't what you are trying to do unconstitutional?"

Senator Lambert's poker face breaks for just a moment before he regains his composure.

"Whether it is or isn't is not our place to decide. If we implement the bill and stall the appeal by Reboot's lawyers in the court system, that will give us plenty of time to acquire the rights to the technology and reverse engineer it before they get the case in front of the Supreme Court. If they win the case, they can have the rights to the technology back. We will have retrieved all the data we need to make our own versions of the technology."

Senator Christianson put down his cigar and brandy. "This maneuver is what we call the double deal. We win either way." They all chuckled together.

Jake adjusted himself in his seat. Being put in this position this fast made him uncomfortable. The secretive nature of the meeting made sense now.

"I don't know if I'm comfortable with voting for a bill I know is unconstitutional."

Lambert leaned in.

"Jake, you don't know the bill is unconstitutional. The courts have not told you if it is or isn't. You are just voting for a bill you want to pass."

"I will ... need to think on this a bit," Jake said.

The group he sat in front of became sober.

"Jake," said Senator Lambert, "we are not your enemy. So, don't make us one. We can offer you almost anything you desire. "You want to be in power the rest of your life? We can provide power. You want to travel the world and stay in five-star hotels? We can provide that. You want fast cars and beautiful women? We can provide those. But, if you go against us, you must be prepared to live a life in squalor. We can provide that too.

"We will give you ample time to make your decision—say by the time we go to session in two weeks. If you don't respond by then, we'll assume your answer is no."

"I understand."

"Good ... good. Please, let me walk you out," said the senator.

The rest of the gentlemen didn't say another word. They watched him leave. When he got to the door and glanced back at them, they all stared back at him with cold eyes. *Welcome to Congress*, he thought to himself.

Sam monitored Jake's tracking signal approaching the outer door. When the door opened, and Jake walked out, Sam left the vehicle to join him. He noticed Jake was a bit pale as they walked back to the vehicle. Sam opened the door to let him get in. He turned back to walk to the driver's door, glancing back at the main entry door.

This should be an interesting debriefing with Reboot, Sam thought. He opened the driver's door and climbed in.

A week later, Sam and Jake traveled back to Texas on a supposed trip for a campaign meeting. During that time, Reboot debriefed them on the details of the meeting Jake had with senator Lambert and the rest of the inner circle back in D.C.

To make certain Reboot's involvement stayed unknown to the government, Sam was given a satellite communicator allowing a triple encrypted signal to be transmitted to a specific satellite Reboot owned and operated, yet no one knew they did.

Reboot board members agreed Jake had no choice but to capitulate to the demands of the group. Reboot didn't care about the vote on the bill, and it didn't matter Jake was forced to vote for the bill. Reboot knew Jake's vote would not be the deciding factor. Besides, if everything went as planned, the government would be overthrown by the time the bill was signed and enforced.

The important thing was Jake adapted well to the political environment and started to be accepted by those in power.

CHAPTER EIGHT

Over the next year, Reboot pulled information from Jake, Sam, and other undercover agents who were in deep cover or tasked with safeguarding specific congressional members.

Thousands of people were tasked with undermining the existing government and putting plans in place to support the coup.

The Inner Circle, as they were dubbed by Reboot, determined the bills submitted by the rest of the congressional members, who voted on what, and how to manipulate the voters. A secret government was subverting the Constitution of the United States within the government—something the signers of the Constitution had tried to prevent from creating in the first place.

The primary task the inner circle's leaders set for Jake to undertake was his first reelection. They told him his ability to run for office would give them a lot of insight into whether his government career would be long or short-lived. They didn't want to waste time on a congressional member who couldn't maintain their position in government.

Reboot felt once Jake was reelected, the time would come to execute the coup d'état and unseat the corruption eating at the country.

Every effort went into getting him reelected. His initial election looked like a cakewalk compared to his reelection. Now with the eyes of power watching, things became more difficult.

The idea was to make his campaign seem like any other. Jake hit the campaign trail like all the other congressional members—beating the drum on the road, advertising to the masses using billions of dollars donated by lobbyists. Reelections and paying back their lobbying partners were all that mattered.

As Jake finished up the fifth week of his campaign, the polls showed him in a comfortable lead against his rival. He represented the twenty-fifth district of Texas.

The district ran from the north side of Austin to the southern end of Dallas. The district followed Highway 35, which spanned between both cities and included a few larger towns.

His retinue had relocated to Waco for the final leg of his campaign before voting was scheduled to open a week later.

Reboot had established a manufacturing site in Waco a few years earlier, which stimulated the economy and increased job security. People started to migrate to Waco after hearing about jobs. This caused a problem with school overpopulation.

In Waco, Jake's speech took place at Tennyson Middle School, where they broke ground on a whole new addition to the school, increasing the grade levels the school could accommodate. It also allowed for much larger population of students.

The school administrators took the idea of university

instruction and added a twist. The school grounds resembled a campus. However, no area of the campus would be separate from the other. Walkways or bridges would access each separate building either above ground or below. Powered walkways would reduce the time required to walk the campus grounds.

The new campus would extend ten stories below ground. This allowed a minimal footprint, excellent efficiency for power and HVAC, and impressive security for the faculty and students.

This was to be the first in a line of upgrades to the school system in Waco. The experiment was to show a location could house many students safely and provide a top education to those attending. Jake's speech focused on the new layout, safety, and of course, the importance of educating the country's future leaders.

Jake finished by answering questions about the new facility from students and faculty alike. The speech streamed live on his campaign site and could be played back at any time to those who missed it.

Once he finished, Sam's team escorted him to his transport to take him to the nearby Ridgewood Country Club for a round of golf.

They pulled into the valet area at the front of the country club. Without the press being present at the country club, it made the location a short getaway for the congressman after a grueling time on the campaign trail. Jake spent weeks giving speech after speech, and prior to the campaign, he spent even more time making marketing spots for ads to be displayed on the web. This was to be a much-needed break.

Sam and his team exited their vehicles and established a small perimeter. Sam directed one of his team members to retrieve the congressman's golf clubs from the back of the SUV. Everyone gave the all-clear, and Sam opened the door for Jake to climb out.

Sam turned to Jake and held out a small duffle bag.

"Jake, here are some clothes to change into. Also, there's a covert PPF in there that will blend in with your clothes better than the one you're wearing now."

The two started walking toward the entrance.

"Thanks. I'm looking forward to this. I wish you'd let your team take the lead while you play a round of golf with me. It feels strange playing alone. Who am I supposed to lose to?"

"I would, but duty calls. We'll make some time after we arrive back in D.C. I want to stay focused here."

"That's what I thought you'd say. I figured I'd try and persuade you anyway."

Sam grinned at him as they walked into the clubhouse.

Jake excused himself to change in the small three-stall restroom. Two guards inspected it and posted a sentry outside.

Sam waited with four other guards in the clubhouse that led out to the course. There were racks of polo shirts, pants, shoes, and every other golfing accessory that could be thought of. The golf course required a seven-figure income to join.

All the amenities a golfer ever wanted were on display. The clubhouse contained a bar with all kinds of imports and domestic beer on tap, a game room for kids, meeting rooms for company meetings, and of course, a top-of-the-line pro shop where those with enough money received lessons from golf pros. Marble and wood flooring rounded out the clubhouse.

Sam browsed some of the racks while they waited for

Jake to change clothes. He realized the clerk behind the counter running the register looked nervous. His body language seemed fine, but his eyes darted around at all the guards. Sam wandered over to one of the guards and whispered to him.

"We checked out all the employees here, didn't we?"

"Yes, sir. Every one of them."

"Keep an eye on the clerk there. He seems ... off."

"Roger that."

A few seconds later, a flamboyant man dressed in a baby blue suit walked in from a nearby office.

"Hey, glad to see everyone ... Oh, wait ... what?"

The guard nearest him had grabbed his arm and motioned for him to raise his arms for a pat-down. Sam walked over to him while security inspected his things.

"You must be the clubhouse manager?"

"Yes! I was just going to welcome the congressman to our humble establishment." He glanced around wide-eyed. "This is a lot of guards for a congressman."

"We like to step up protection during reelection campaigns," replied Sam

The guard finished patting him down and nodded to Sam.

"You have a lot of impressive merchandise here. The congressman may end up buying some of it, but you'll need to wait to peddle your merchandise until the end of his round of golf. We have a schedule to keep."

"I understand, er—what was your name?"

"My name is Sam."

"Glad to meet you, Sam. My name is Cory," said the clubhouse manager.

About that time, another employee came in from the caddy area.

"Perfect timing, Jona," Cory said.

Cory motioned to Jona while looking at Sam. "He would have been Mr. Tillman's caddy for the day."

The comment took Sam a second to contemplate, but that was all the time Cory needed. The alarm in Sam's mind had started to go off when Cory kicked out with his foot and snapped the guard's leg at the knee that had just patted him down. He pivoted off his leg and spun to land a side kick straight into Sam's chest, sending him flying backward. The guard screamed as his bone broke the skin in his leg and protruded from the wound like a broken twig from a tree.

The attendant behind the counter pulled out a BB-10 slug launcher. The weapon was an inexpensive firearm designed to fire a slow PPF penetrating round. The guard nearest Jona had a bead on the attendant, but Jona interrupted him before he fired off a shot by smacking him in the face with a nine iron he had snatched from a nearby display rack. The guard's tranquilizer rounds shot into the ceiling.

Sam had pulled his sword with his right hand while he detached his ST-115 from his hip that fired two four-inch rounds designed to penetrate PPFs. The two men fired at each other at the same time. Sam's customized PPF affected the trajectory of the slug enough that it missed his head by inches.

He fired both his rounds. The guy behind the counter wasn't wearing a PPF and took both straight into his chest. He crumpled to the ground behind the counter.

Cory used the time to locate a hidden panel and pull open a weapons cache. He pulled out a grenade and popped the pin. He tossed it in the direction of the two guards who had come running in from the hallway from the bathroom where Jake was changing. The blast blew them right off their feet, killing them.

Sam was knocked toward Jona by the blast. Jona had finished beating the other guard unconscious with the nine iron and had turned toward Sam the instant the grenade went off.

Sam threw his empty ST-115 at Jona and went into a roll. Jona dodged the weapon and came at Sam with the golf club. Sam came out of the roll with his sword up in a defensive posture. The sword's edge cut through the graphite club shaft as if it were tin foil.

Jona stared at the end of the club in disbelief. That was long enough for Sam to go on the offensive. Jona raised what was left of his club up to block Sam's strike with his sword. Sam's sword went straight through the shaft of the club, through Jona's shoulder, and into his chest, causing catastrophic damage to his lungs and heart. He died before he hit the floor.

As Jona hit the ground, a knife struck Sam deep into the meat of his left shoulder. He turned in time to parry away the second knife Cory threw at him. Sam yanked the knife out of his shoulder as he watched Cory pull a sword out of a scabbard he pulled from the secret stash in the wall.

That left Sam and Cory in a standoff. Sam's offhand shoulder was bleeding a bit. He didn't feel like anything important was damaged. He cursed himself for being caught off-guard by the ambush.

Sam worried about Jake, not knowing if others had come in to snatch his friend while he dealt with this group.

Smoldering embers from the grenade littered the area. Smoke filled the room with a light haze as the two men faced off.

Sam could identify the molecular sword by the handle

where the hi-output battery was stored. The edge of the blade would not give in to Sam's strikes like the golf club did.

"Why are you here?" Sam asked.

"I was sent to kill you," Cory replied simply.

Cory took a quick step forward and swiped up and to the left with his sword. Sam countered with parries from his own sword. Sam spun in low after the high parry in a counter. Cory made the mistake of using his sword to block the strike instead of jumping over it. This left him open to a spinning kick that Sam followed up with. It connected with the side of Cory's face and staggered him. Sam continued with two more slashes at his feet. Cory blocked one and tried to jump over the second. However, he was off-balance from the kick to the head and didn't get one of his feet up in time. His left foot came off as Sam's sword sliced through it.

The shock of his foot being amputated hit Cory as Sam twirled around and took off his head. His body plopped to the floor. Blood still pumped from the headless corpse as the heart continued to pump for a few more beats before stopping.

Sam gave a quick inspection of the destroyed room. He picked up his ST-115 and ran for the bathroom where Jake had been changing.

Silent as a mouse, he made his way to the door and pushed on it with a finger. When the door failed to open, he knew it was locked. He gripped his sword with both hands and made two quick slashes around the edge of the door. He lashed out with a front kick, and the door splintered apart.

At first glance, the bathroom looked empty. Three stalls and two urinals filled the room.

"Jake? Jake, you there?" Sam yelled out.

Jake's head popped out of the top of the farthest stall from the door.

"Sam? Thank God!" Jake replied as he jumped out of the stall wearing his golfing gear.

"Is your PPF on?" Sam asked, pointing at Jake's belt he gave him earlier.

"Yes."

"Awesome, let's move."

Jake nodded.

"Stay close to me. We're going to move for the SUV."

"Understood," Jake replied.

Sam slipped two more rounds back into this ST-115 and slammed the barrels shut. He handed it to Jake.

"Take this."

Jake nodded.

He hefted the weapon with a familiar hand. It had been one of his favorite weapons for short to medium-range combat when he was in training.

Jake kept one hand on Sam's back as the two moved out of the building and to the SUV. When they opened the doors to go outside, they were greeted with a chorus of bird song and sunlight.

From all appearances, it was just another day on the course. The SUV sat fifteen feet from the entrance. Sam pulled a remote from his pocket and hit a button. The SUV started up and sat there humming. He pressed another button, and the passenger door popped open. He sheathed his sword.

"On three, we are going for the passenger side. I'll scramble into the driver's seat. You get in as fast as possible."

"Understood, Sam."

"OK ... three!"

The two of them sprinted the fifteen feet to the SUV

and dove in. Sam got into the driver's seat with a quick glance into the back to make certain no one hid there in ambush. He threw the vehicle in gear and slammed the accelerator pedal, smoking all four tires on the pavement as it lurched forward.

The door slammed shut as Sam pressed a button on the driver's door. Jake slipped into the back to grab a first aid kit. He had discovered Sam was bleeding from his shoulder when they made their way out.

"Birdcage, this is Sierra Juliet. We have a condition red. Operatives are down on site. Package is secure and moving to extraction Charlie. Do you copy?"

"We copy you, Sierra Juliet. Confirmed condition red, and pickup will be waiting at extraction Charlie. We are tracking your location. You are two mikes out. Confirm."

"Roger that, Birdcage. Two mikes out."

Sam finished up as Jake crawled back into the passenger seat with the first aid kit. He started dressing Sam's wound as Sam made his way back to the extraction.

"So, are you going to tell me what the hell happened back there? I go to change and almost get knocked off my feet by an explosion. So, I locked the door and hid in the stall."

"I am not certain myself. These guys passed the background check. They attacked us after someone named Cory introduced himself."

Sam paused a moment.

"I don't know if the other guys are alive are not. I hope they are, but Ron and Brad both took a grenade point-blank. Rob was a bit further away but had his leg shattered, and Tom was knocked out by a golf club to the face. I took a hit from a throwing knife by the supposed clubhouse manager calling himself Cory. He and his team are dead, but I don't know if there were more. The strange thing is I don't think they came there for you. They appeared to be there for me."

"You?"

"Yeah, as strange as that sounds. Cory, their lead man, said he came there to kill me."

"That doesn't make any sense," Jake said. "What would killing you do for anyone other than remove you from my security detail? I mean, is it that much of a rewarding position that someone would kill you so they could get it?" he joked.

"Maybe the goal was to slip in close to you by eliminating me. Though, why try and take me out when we're together? There are plenty of other ways to go about removing me. I mean, I expected trouble while on the job, but if I'm at the corner shop getting groceries, wouldn't that be a better time to try and kill me?"

"I'm just glad you made it through, and we both got out in one piece. Well, mostly," he said while he wrapped Sam's shoulder with gauze.

They drove on without further incident to the extraction and were taken to a secure location for debriefing while the site was cleared and investigated.

No further evidence presented itself in the aftermath of the attack. Reboot assumed the leaders of congress who Jake had met with were behind the attack.

For reasons unknown to Reboot, they tried removing Sam from leading Jake's security detail. Either they thought he signified a threat, or he got in their way when it came to grooming Jake. Reboot sat that aside as an interesting side note while they kept pressing forward with the coup.

Jake's reelection went off without a hitch. He won by a fair margin, nothing too grandiose. They didn't want to raise suspicions by routing the other candidate.

Reboot had positioned all their pieces as well as they could. Key military members, congressional members, lobbyists, police, and the rest of their agents were ready for their word.

Seven months after Jake's reelection on June 2, 2143, Reboot informed Jake and Sam in a debriefing that the weekend before July 4 would be the start of the next stage.

Key agents were going to eliminate all members who had been deemed hazardous to the future of the country. Inside military personnel would hold back the military reaction long enough for the President, Vice President, and Speaker of the House to be removed from office.

Not long after the briefing, Sam noticed Jake's habits started to change. He became more nervous by the day and had started breaking eye contact with Sam when they were talking. He had stopped conversing with him like he'd used to.

An emotional barrier started to appear between them. He started arriving late for meetings and gave poor excuses when Sam attempted to broach the subject with him. Sam could understand the amount of stress he was under due to their mission, but they were always open with each other when alone. Jake seemed to be preventing those situations as much as possible.

Within a couple of weeks, they had progressed to the execution of the biggest coup in history. Sam reported to Reboot in secret that he no longer trusted Jake. Reboot insisted Jake reported to them everything they required in his personal briefings. They suggested Jake was reacting to the extreme stress of being undercover for that amount of time. His life was on the line, after all.

Sam was reluctant to agree, but he gave way nonetheless.

CHAPTER NINE

The month of June neared its end. The weather turned out to be quite pleasant and was forecast to be the same well into July.

July fourth landed on a Wednesday that year. That made the entire week a holiday for the capital city. Red, white, and blue decorations popped out of every nook and cranny throughout the city. No expense was spared to make the capital look like it came out of a pre-war time era. People enjoyed the monuments and parades as a welcome distraction to life outside the city.

With all the congressional members attending the festivities that week, Reboot operatives would have no problem reaching their targets.

A courier notified Sam and Jake that Reboot wanted an emergency meeting two days before the start of the coup. Jake found an excuse to leave his office for the afternoon, and the two drove out of town to a small coastal town called Fairhaven.

Months before, Sam purchased a rundown beach house there under an alias. Fairhaven was a ghost town, like hundreds of other small towns around the country since the pandemic.

Trees, bushes, grass, and weeds had all been given free rein of the town after it was abandoned. Vines grew up the old lamp posts and hid their rusted trunks in a carpet of green. Ivy grew across whole sections of the sidewalks and roads in the area.

Birds took advantage of the many available roosts to build their summer homes. Even the national bird started appearing around the capital again. Many of the Bald Eagles could be seen flying over the capital on their way to hunting grounds on the Potomac River.

The town had been a middle-upper class location for people who wanted a summer beach house away from their primary home.

Sam's place was listed on the bank's notary as reclaimed, the same as millions of other homes throughout the country. It made a secluded spot for a communications outpost. Sam would go there often to get away from the city. He used the time to clear his head and relax. Jake didn't have the same luxury.

The drive out there was uneventful. Jake didn't say much at all to Sam on the way out. Sam tried to start up conversations during the drive. However, Jake kept his answers short, staring out the window at the brush passing by or watching critters scramble out of the brush they drove through.

The country's capital did its best to hide the fact most cities appeared as run-down husks of their former selves. The city remained clean and sparkling. Once someone reached the city limits, the cityscape became a much drearier picture.

Nature was hard at work, taking back what humanity

once took from it. Where a bustling suburb once existed, a sea of vacant homes covered in green replaced the tidy lawns and brightly colored homes.

Trees grew out of gutters filled with debris. Vines and other plants grew up the sides of the houses. Yards that used to be well-groomed grass and flowers decades ago grew into thickets a person could not pass through.

The department of transportation still maintained the primary interstate highways and local primary avenues of travel in the cities. The side roads were left for nature to overtake—any travel not utilizing those primary routes of travel required SUVs and other off-road vehicles to traverse.

The black SUV pulled into the drive of Sam's house. The scent of the sea filled their nostrils as they got out of the armored SUV and walked into the home. Sam disabled his security system and did a quick check of the premises.

He invested a lot of money to clean up and restore the home, the lots to either side acting as a reminder of what his home looked like when he had first purchased it.

Sam guided Jake into the secret communication room he created that only allowed his tablet to communicate to the stationary satellite located hundreds of miles above them.

Sam got the connection up while Jake sat there with a blank expression on his face. Robert Bartram appeared on the screen and looked over the two agents sitting before him.

Sam appeared confident and stern. His pale blue eyes remained vibrant and intense, but his expression became wary when his eyes fell upon Jake.

Sam had taken advantage of not having to be in a suit, opting for shorts with cargo pockets and a short-sleeved polo shirt instead. His ST-115 and sword remained within reach.

Jake's puffy, bloodshot eyes made him look ill in comparison. His expression seemed absent. Both men needed some sun from the neck down, the price for wearing suits all day. The designer jeans and t-shirt Jake wore did little to hide his demeanor.

After looking over the men, Robert spoke.

"Alright, men. I know time dictates we make this quick. This meeting would not have been requested if the situation was not important."

Jake and Sam perked up when they realized the meeting brought bad news.

"We are losing contact with operatives in the field and with contacts within the military and government sectors at an alarming rate. Something bad is going on. We believe someone leaked information to the government concerning our plan. We appear to have a mole, gentleman. Until we find more information, we are holding off on the operation until further notice."

"When might we know more?" a worried expression flashing across his face.

"I have no idea. We are notifying the remaining contacts we are able to get in touch with. You two are one of the highest priorities."

"I think this may cost us some months at the very least. Some of these contacts are important to the later stages of the coup. We cannot hope to succeed without them."

"What do we do now?" Sam asked

"Keep going along with your current assignment as if nothing changed. Jake, that means preparation for another reelection campaign."

Jake visibly shuddered at the comment.

"Sam, we are thinking of pulling you from the detail and putting you someplace else. Jake should be able to use the regular security teams now that he's familiar with the system and who to trust. Would you be comfortable with that decision, Jake?"

"I'm fine with that. I'll be alright without Sam. I've made plenty of contacts here and earned their trust."

"How about you, Sam?"

"Fine ..."

Sam's eyes went to Jake. Jake just sat there, ignoring his stare.

"I'll report your thoughts on this, and we'll update you both as soon as possible. Jake, get some rest you look beat and don't worry, we'll get to the bottom of this.".

After they said their goodbyes, Sam and Jake locked up the house and headed back to D.C. Sam sat in the driver's seat, seething in frustration at Jake's response to pull him from the detail. Jake stared out the window in silence. A couple of miles down the road, Sam found a little area not too overgrown to pull over and brought the armored SUV to a stop, throwing the gear in park. Jake kept looking out the window at the old homes slowly succumbing to the ravages of nature and time.

"We better get back," Jake said.

"What in the hell is wrong with you?"

Jake said nothing. Sam reached over and grabbed his arm.

"Jake! Look at me!"

Jake whipped his head over at him.

"Let go of my arm, Sam," he said in a calm voice.

"First you tell me what the hell is going on with you. We used to be like brothers, and now you treat me like we never met before—hell, even then you were more talkative! Is the job getting to you?"

"No."

"Have I done something to piss you off?"

Jake looked down at his lap. "No."

"Then what is the problem?"

"The time has come for you to leave, Sam. You aren't needed here anymore. There are more important things you can be doing than protecting me now. This cloak and dagger stuff we do is too hard. I need to be in character 100% of the time. I can't afford to be your friend and congressman Jake Tillman. I can only be Jake. If you are my friend and brother as you say, you'll stop leading the security detail."

"If that is what it takes to be your friend, then so be it."

Jake replied quietly, "It is safer for you."

Sam gave Jake a level look. "What does that mean?"

"I don't know if I can protect you any longer."

"OK, so what does *that* mean?"

"Just drive. We need to head back."

Sam sighed and let go of Jake's arm. Jake turned back to stare out the window. Sam sat there a moment, looking at his friend. A moment of grief washed over his face. Then, he decided to embrace the problem and move on.

"Very well, I'll drop you off with the security detail."

Jake didn't respond. Sam paused for a moment and glanced back over at Jake, and then he drove on.

CHAPTER TEN

Sam had been off the detail now for a few days. July 3 brought major changes to the city in preparation for the July 4 celebration. Sam sat on his couch in the small apartment he rented in the city when a knock came at the door.

He grabbed his sword and clicked on his PPF as he went to the door. He checked the monitor to see who rang—it appeared to be a courier.

He cracked the door and peeked out.

"Can I help you?" he asked the young man at the door.

"I have a message for a Sam Miller," he replied.

"That's me."

The courier handed him a small envelope with his name printed on the front, then left. Sam closed the door and peered at the envelope. He held the thin white envelope to the light. A thick piece of paper with a single line of text appearing on it was contained within. Sam couldn't read the words through the envelope, so he tore it open. A single sentence was written across the piece of paper.

Meet me at this address tonight at midnight. There is an emergency meeting. -Jake

The method of communication Jake used gave Sam an uneasy feeling. The two of them used code words via electronic communication. They would never use something as open as this.

He remembered the last conversation he had with Jake in the truck on the way back the other day. Jake mentioned it was safer for Sam not to be in charge of his detail any longer and that he couldn't protect Sam. Maybe this was the safest way of communicating with Sam now.

Sam looked up the address. It was an abandoned post office in a derelict part of town north of D.C. This might be his orders for his next mission, so Sam packed up his important belongings.

He dressed in his tactical gear and would approach the site with stealth in case things were not as they should be. Jake knew Sam never went anywhere without his ST-115 and molecular sword. Carrying weapons should not come off as abnormal.

Sam arrived at the location around eleven to survey the scene. The post office stood alone in an old parking lot. Rusted deposit boxes, tall grass and weeds, and overgrown trees dominated the location. No lights were on in the building, and no vehicles were present at the location either.

He staked out the location from the nearest building across the street from the post office. The building resembled some sort of facility where people would climb up the walls. Handholds were strewn over a fake rock face, reaching almost to the ceiling. Old pulleys hung from there, and at the base of the wall, old foam padding shredded by rodents appeared to be used for safety purposes. The stench of years of rodent urine and feces permeated the air.

Sam used a small pair of thermal binoculars to scan for any signs of an ambush. He didn't find anything other than a stray dog walking through the area. He glanced at his watch. The time indicated ten after eleven. Sam thought about the scene for a moment and then decided he would sneak into the post office and wait for Jake.

He sprinted the short distance from the office building he was in to the side door of the post office. The door had been pried open long ago. Something had bent the steel exterior door at the latch and prevented it from closing. Sam squeezed through the opening and slipped inside the dark interior.

His sensitive eyes adjusted in the darkness.

Old carts used for hauling packages sat empty, and disintegrated box debris littered the area. The place had been picked clean of anything useful long ago. Sam moved from room to room to make certain he found nothing out of place.

He latched and secured the two garage doors in the back docking bay. He took some broken glass he found and scattered it at the back door that remained locked and the side door he came through. He positioned himself behind the front counter where patrons stood to receive or ship their packages. Once he was comfortable, he waited for midnight.

Headlights appeared on the road in front of the post office about forty minutes later. The familiar whine of a black armored SUV pulled up in front of the officee, and the driver's door opened and closed after the vehicle turned off.

Sam monitored Jake's movements as he walked towards

the front of the post office with his communication pad they used to speak with Reboot. No other security detail members accompanied him.

Jake walked in through the front turn-style door. The door shrieked over having to move the old, rusted bearings supporting the door. Once he progressed through the vestibule, he turned on a small flashlight and made his way in through the front area.

Sam stayed hidden for the moment to make certain Jake wasn't up to anything. He wanted to trust his friend. However, things changed too much for him to do that any longer. Sam longed for the lie-detecting chairs he and Jake had sat in during their briefings at Reboot.

Jake pulled over an old stool, took out a handkerchief, and laid it over the filthy foam seat of the stool. He sat down and laid the tablet on the counter, then propped it up with a built-in leg on the back of the device. Jake pressed a button to turn the tablet on and waited.

When midnight came, Sam stood from behind the counter. Jake noticed the movement and flipped on his flashlight, pointing the small light at Sam.

"How long have you been here?"

"I arrived just before you."

"Sure you did," Jake replied sarcastically.

Jake wore a dark suit with a white shirt and some dark red tie that mixed in with the limited light of the post office surroundings. He seemed legit to Sam, but Sam also knew Jake was trained the same as he and that he might be trying to disarm him by being his normal self.

"How did Reboot contact you?" Sam asked.

"The same way I contacted you, by courier."

"Why didn't they send a communication to me?"

"I don't know. Ask them when they come on."

Jake typed in a passcode on his tablet, and they waited for the program to go through its encryption process, bouncing the signal through multiple relays to prevent monitoring.

The screen came up, and an empty office chair sat in front of the terminal.

"Are we on? Sorry, gentlemen. Give me one moment," said a strange voice.

Bryan Lambert, the senator from Texas, appeared on the screen and sat down in the chair with a sigh. Sam muted the screen.

"Isn't this the guy you met with in D.C. when you were first elected?"

"Yes," replied Jake without a hint of emotion.

Sam's brow furrowed.

"Why is he sitting at our Reboot location?"

"I don't know. Maybe we can ask him and find out?"

"Sorry, gentlemen, I can't hear you," said congressman Lambert.

Jake unmuted the tablet.

"We are here, Mr. Lambert," Sam replied.

"That's congressman or Senator Lambert, but no matter. It is an honor to finally meet you, Sam. Jake told me a lot about you. I doubt the same can be said for myself. I know this may come as a surprise. You were expecting someone else, no?"

The senator sat back in the chair with his arms behind his head.

Sam frowned and glared over at Jake.

"I have to say. It took some enormous balls to try and overthrow the U.S. government. No entity has ever even attempted such an endeavor."

"Unless you count the Civil War," Sam snapped back.

The senator continued unabated. "I give your team an A for effort. Unfortunately for you, we picked up on your little scheme, and as you can see, we did something about it. But let me take a step back. I don't want to jump ahead of myself.

"My little group of friends realized something was different about you two the moment we saw you. The saying, it takes one to know one, strikes true here. It took significant time to figure out what your exact agenda entailed. At first, we thought you were some attempt by a media institution to expose us. Once we confirmed with our contacts that neither of you was affiliated with the press, it became clear you hid a much darker agenda.

"Then we focused on assassination. We assumed you were going for the President. Which in all honesty, we wouldn't care a bit about. We couldn't know for certain until we broke one of you. Lucky for us, Jake was willing to exchange evidence in place for a full pardon and entry into our little club here. Of course, we needed to inject a smart bomb into his chest to make certain he would follow up on the deal."

A scowl crossed Sam's face as he learned Jake was the mole. His hand twitched as it inched towards the sword sheathed on his back Jake continued to stare at the screen with the same blank expression he had since he arrived.

"Don't blame Jake, Sam. We can be very persuasive when we want to. People do crazy things to prevent torture or death. We are the ones who made him send the group to the golf course to kill you. We underestimated your skill. However, we learned a lot about your abilities in the recordings we have of the incident.

"That brings us to tonight. I'm sitting in the chair of whomever you spoke to on a regular basis–this Robert fellow, or is it Laura? Ah, no matter. In time, we'll determine that information. You, Sam, have the honor of being the first execution for treason in over a century. Of course, you won't be the last. Tonight will be a cleansing of sorts. By the time people wake up to celebrate the birth of our independence, Reboot will cease to exist.

"That leaves a big decision up to you. A SEAL team will be descending on your location in a few moments. I see three options for you. One, you can kill Jake for his betrayal and then die. Two, you can prevent him from leaving, and you both die together. Three, you can let him leave now, you die alone as a traitor, and he comes to work for us. But, before we delve to the meat of the matter, do you have any questions?"

Sam turned to Jake. "Why?"

Jake made eye contact with Sam for the first time since the senator came on the screen.

"Sam, you have no idea what it is like having to be me every minute of every day. You found opportunities to escape the city without reason and clear your mind. For me, the days have been torture. At a certain point, I couldn't take it any longer. It wasn't like I could tell our handlers that I was spent and had nothing left. It would have screwed their plans up to the point of having to start over, and I am not going to let them do this to another person. I'd rather see them fall. Reboot brought me more power than I ever imagined, and I'm thankful for that. But not to the point of sacrificing my life so they can kill thousands in an attempted coup.

"Giving up my best friend is the hardest thing I've ever had to do. It's why I turned my back on you. That was my grieving process ... I let you go. Now you need to let me go. Let me walk out of here and live my life."

"You think the thousands who might die in this coup wouldn't have the blood of thousands or even millions on their hands already?"

Sam pointed at the senator on the screen. "This jackass and all those like him killed many times that in their rise to power and their scramble to protect it. They care nothing of those they trample; you know this! My parents were murdered for nothing more than to protect the power of money. That's why I was orphaned. Reboot gave me the chance to right a wrong. If I die, I die."

Sam took a deep breath, trying to control his anger. "You better leave, Jake, and pray I don't live. Because if I do, the next time we meet, I will kill you for what you've done."

"I understand, Sam."

Jake disconnected the tablet and took it with him as he walked out, leaving Sam to stand there alone in the darkness. Sam's rage grew moment by moment, and soon, the whine of electric motors could be heard as Jake's armored SUV drove away from the post office.

Sam prayed. "Lord, please hear my plea. I apologize for any sins done in my life and any I'm about to do in the next few moments. If I live through this, I plan on hunting down all those responsible for the deaths of so many. I want nothing but to help those in need, and I feel this country is in need. Give me the stealth, the speed, and the strength to accomplish my task. I will be forever grateful for your presence and your guidance. In Jesus's name, I pray, amen."

Sam opened his eyes. Now that the bright screen of the tablet was gone, Sam could make out the interior of the post office again. The SEAL team would be at his location soon.

He expected they might use snipers to cover any exits of the building, which meant leaving that way was not an option. He remembered seeing a manhole in the back utility closet where the utilities ran into the building. The manhole may lead into a sewer access tunnel he could use to escape from. With his luck, it wouldn't lead anywhere. It was his only shot.

He moved from his location to the hallway he came in from. He peeked down the hallway towards the door. He didn't recall hearing any glass being disturbed earlier but the door was wedged open now. The Seal team could have slipped in during the senator's drawn-out speech.

Sam pulled a flash-bang and a smoke grenade from his belt. He threw the smoke grenade so that it landed just outside of the door and waited ten seconds for the smoke to thicken. He followed up with the flash-bang. That would disorient anyone's scopes used to peer down the hallway.

The flash-bang went off, and grunts emanated from the location. He took a few steps back and sprinted across the hallway towards the docking bay.

A couple of heavy sub-sonic rounds flew down the hallway past him and lodged in the end of the hallway, exploding a second later. The sound of glass crunching could be heard from the other door a moment after that. Sam sprinted to the utility room after dropping his last smoke grenade. A padlock secured the top of the access cover. He would give away his position by cutting the lock with his sword, but he had no choice.

He pulled the powered blade from its scabbard and gave the lock a hard slash. A loud clang came in response to the lock breaking free.

He pulled the lid free from the entry and breathed a sigh of relief. An opening appeared, leading down into the sewer system. Sam scrambled down the hole, pulling the cap back into place behind him.

He figured he had a couple of minutes or less before

they realized he escaped into the sewer. It would not take them long to pull up a map of the sewer system he didn't have access to. They had the advantage.

He moved as fast as he could down the dry sewer tunnel for as far as he could go before coming to an intersection. East would take him back in the direction of his apartment. West would lead towards the Potomac River.

The post office was located on the edge of an abandoned Air Force Base that closed at the end of the pandemic era.

The base processed the dead and dying. Once the plague was over, crematoriums ran day and night to burn the bodies up. It shut down after the furnaces finished with processing the dead. The base bordered the river's east bank if his perfect memory served him.

Sam decided heading towards the river was his best option. They would be searching his apartment for any information on Reboot, so he abandoned any plan for returning there. The river was his best chance.

His first obstacle was an iron grate put in place to prevent infiltration into high-security parts of the base. His sword made short work of the old iron bars standing in his way. He couldn't hide the fact he passed that way, and if the team pursuing him were Seals, they would split up into two teams upon discovering he went underground. It wouldn't take long to determine which way he went.

Sam traveled for some minutes headed west. He came to a dead-end that fed into a leg of the sewer drain-off from the base. Sam looked above him. A bunch of inlet pipes led into this section. That meant a building was overhead or near his location. He backtracked to a ladder about forty feet back the way he came.

Sam gave the cover a hard push, and it didn't do much

but wiggle. He had no room to hack at the iron plate with his sword, and he didn't have any explosives to blow the iron cover open. He pushed on it with all the strength he had in his arms, and the manhole opened enough to see a bit of light for a moment before the cover slammed shut on him. Something heavy had to be resting on the manhole cover. The pungent air started making him feel claustrophobic and a bit sick to his stomach. He swallowed back the panic rising in his throat and focused.

He thought about backtracking even further to the next manhole when he saw a light flicker from the way he came. The team had caught up to him. This was not a place to try and fight off some Navy SEALs. His sensitivity to light would make him blind if one of them shined a flashlight in his face.

Sam decided to change his tactics. The problem was leverage. He needed to involve his legs in the task. Gritting his teeth, Sam climbed back up to the manhole cover and flipped upside down on the ladder. He clung to the ladder like a weightlifter doing a squat. He used the ladder as a brace and got a firm hold of a rung near his waste, planting his feet on the manhole cover. Once he obtained a firm grip, he gave a test push with his legs, and the plate shifted. Sam grunted and pressed his legs against the cover with all his strength, and the cover rose about six inches. He grabbed a rung closer to the top and pulled himself back into a crouch while keeping the cover at the same level. Sweat dripped off his face from the effort.

Still upside down, Sam craned his neck to see how far down the tunnel the SEAL team had made it. Their footsteps echoed off the concrete drainpipe as they approached. The light from their flashlights were almost within range for them to see him now.

He gave another hard push, and the cover gave way with a crash. The sound boomed down the shaft and into the tunnel. The team's footsteps quickened as they rushed toward the sound. He pulled himself out of the tunnel just as they made contact.

He slipped the cover back over the manhole and did a quick survey of his surroundings. He found himself in a multi-stall garage, appearing to be a maintenance depot. Vehicle repair bays lined either side of the concrete walls for a hundred feet.

A massive, old steel shelf with a bunch of junk parts on it rested on its side next to the manhole cover. That would take too long to put back in place. Sam scanned the area and found what would give him a bit of time. An old escort motorcycle lay near him with cobwebs all over it. He raced over and grabbed the motorcycle by the rusted front wheel. The tires were flat, and the engine resembled a ball of rust, but he needed something heavy, and the motorcycle would do.

Sam pushed and dragged it over to the manhole as fast as he could. He got back in time to see the manhole cover raising a few inches. He let go of the bike and jumped on the manhole cover with all his weight. Some cursing could be heard as the lid slammed back down.

Sam reached over and dragged the bike to him. He slid himself off the cover and pulled the motorcycle on top of it with the engine centered over the cover.

The bike must have been used to run escort service for dignitaries decades ago. The rusted hulk gave him the precious time he needed to attempt his escape.

An exit door between two garage doors with a broken window rattled in its frame from the wind. He ran to the creaking door and peeked out.

He could smell the moisture—the river lay nearby. Sam glanced back at the motorcycle. The SEAL team wouldn't be able to track him once he entered the water. He grabbed the doorknob and gave it a twist. The rusty innards groaned at him, but they still remembered their job and released the latch.

Plenty of hours of darkness remained before the sky would start to show signs of dawn. He ran west away from the old building and into the darkness beyond. Sam could hear the river. He climbed a flood wall embankment, and the river greeted him like some massive black snake working its way past.

Sam flinched as an explosion echoed back from where he came. The team must have blown the cover off with an explosive. If they split up where he thought they might, the other half of the team could be nearby. He would be easy to spot with infra-red.

Sam looked back at the freezing river. The water would help mask that. He dashed down the bank and slowed to a walk as he slipped into the Potomac with as little noise as possible. The cool water felt good against his sweaty skin.

Sam realized after entering the water that an airport existed on the other bank of the river. It appeared to be Ronald Reagan Washington Airport based on what he'd memorized off maps for the area. Lights streamed from the airport grounds. That much wide-open ground would seal his fate if he tried coming onshore there. He would need to let the current take him downstream to a safer crossing.

He remembered the bridge for Interstate 495 was a couple of miles south of the airport. Making landfall there would keep him within too close of a search radius for drones. He would let the current take him far downstream. The river went due south and then snaked west.

The Potomac River emptied out into the sea to the east. As long as he headed south or west, he would stay in the water near shore. This would allow him to put some distance between him and the search area while saving energy. He hoped any drones scanning for him would fail to search near the water.

He passed under the bridge about twenty to thirty minutes after entering the water. The current moved him slower than he expected—he figured near the speed of a slow walk. He would never progress any faster than that here in the river delta area. The water was close to the ocean, and the area was near sea level. The Potomac wasn't known for anything other than a safe traveling route for barges.

After he traveled with the current south, west, and then south again, he came upon an immense wooded area where no remains of buildings were located on the west side of the river. The area may have been a wildlife refuge from years past. Nothing else could explain the reason no buildings remained near it this close to the capital. He pulled his waterlogged body out of the water and up into the brush.

The lack of people meant an easier time not getting reported to authorities. The real threat was the unmanned drones the government used to hunt down criminals and maintain order. They would be out in force soon, if not already. He was safe for the moment, but he dared not rest. This scenario made all the hours of cardio he put in every week worth every minute.

His thoughts turned to Jake as he ran through the night. He couldn't believe Jake turned on Reboot. If the senator was at the location Sam was debriefed out of, then the government must have moved on Reboot enforce. He would need to locate a safe house and establish communication to any remaining Reboot command. He knew of one not far from D.C.

However, Jake knew of the location as well. Arriving at that safe house would be a mistake. He would make his way to a train station and jump on a cargo train.

He knew of a couple of other remote safe houses in Ohio he would prefer trying before getting jumped by a SEAL team at the nearest one in Virginia.

His foster parents taught him all about survival well before he trained at Reboot. They used to camp for weeks at a time in the Texas countryside, though they had tents, cooking stoves, and food. He had his sword, ST-115, PPF, and the clothes on his back.

His thoughts went to his friends and parents, and he hoped they were able to escape before the government moved in. He shook those ill feelings aside.

The best place to jump a train would be Manassas Station. Freight trains often stopped at that station before heading into West Virginia. Train systems were the fastest way to travel and required much less maintenance than the road systems. Most roads were impassable without an all-wheel-drive vehicle.

The rail systems were all maglev-based systems. The freight trains didn't run near the speeds of a passenger train due to sheer weight and required energy, but they still managed to approach two hundred miles per hour on the long, straight sections. They remained much slower in the mountainous area where he would be boarding.

His goal was to make his way to Parkersburg, West Virginia, along the border of Ohio on the Ohio River. The town contained a switching station and remained within fifteen miles of the safe house he wanted to scope out.

Sam's photographic memory allowed him to memorize

all the safe houses Reboot maintained all over the world within a few minutes of studying the map. He was confident Jake would not have access to such a random safe house location, nor would they think he would go so far away before trying one.

Sam approached the Manassas station from the south. The switching station was to the west of the boarding station. He assumed by this point that he was one of the topmost wanted in the country. Facial recognition would pick him up seconds after reaching the station floor. Any public place was off-limits now.

Sam let out a huff and looked around. He would have to find a way on a train without going through the station.

CHAPTER ELEVEN

Daylight broke the horizon as Sam took account of his gear. He'd dumped all but the most essential gear as he'd escaped into the Potomac River. That left him with his black t-shirt, black on grey camouflage cargo pants, his sword, and ST-115. He looked mighty suspicious. Other than going naked, there wasn't much he could do about that now.

Manassas Station was far enough outside of the capital that not many people lived there. Most of the trains that stopped here contained cargo. Getting on a maglev train was no easy thing, but it was his best option.

Sam found a line of cars loaded and waiting for a train engine to be hooked up. He waited about a football field's length away from the rows of train cars. About midday, he saw an engine being backed into place and secured to the line of about twenty rail cars.

Sam didn't know much about trains, so he would need to improvise. He picked a black box car—the color would help him blend in when he jumped on it. He prayed the black train car contained enough room for him once he boarded. It would be difficult to hold on once the train started getting up to speed.

The train's engineers finished their checks of the cars to make certain everything was connected and secured. The three men jumped on a little electric cart and made their way back towards the station. A moment later, the conductor climbed aboard the massive engine, which was about fifty feet long with a pulsing light going down the center of each side of the train.

Sam's foster father had explained the basics of a maglev system to him many years before. A massive amount of electrical energy was needed to power a maglev system. The rail took on the opposite polarity the engine powered the train cars with. Once the strength of the opposing polarities reached a high enough point, the train cars would levitate away from the rail. That is why they could move so fast; there were no moving parts to the train system. Once the power level reached the peak for the train, the entire train would be repelled by the rails and hover over them by a couple of inches. The longer the length of the train of cars the more engines needed to be attached to power them. The more mass the more energy needed to move it. Sam just needed to remember to stay away from anything labeled high voltage.

Once the conductor disappeared into the engine compartment, Sam decided the time was right to sneak closer. He would have a few moments to jump on the train car before it outpaced him.

The pulsing in the main engine started moving faster. All at once, the entire train lifted off the tracks. Sam slipped out from between two cars on another track and sprinted towards his target car, glancing overhead for drones as he went.

The train lurched forward and started gaining speed as Sam grabbed the handle for the sliding door. He investigated the locking mechanism to see how he could bypass it, but no physical lock could be identified. Sam assumed the lock must be magnetic as well. Wires needed to be run to the door to keep the magnet powered and prevent the door from sliding open.

He pulled his sword from its sheath on his back and hacked at the car wall behind where he thought the wires might be. After a couple of poor attempts to land a blow, he struck home with a strong slash and sliced into the side of the car. The door sizzled as it slid free.

Little room remained in the train car. The train picked up speed to the point where Sam was being buffeted by the wind. He managed to maneuver up and over a massive storage container that sat right at the door.

Once he was over the container at the cargo door, he found enough space to squeeze into and sit. After kicking at some bits of cargo, he was able to wedge the entry shut.

He made himself comfortable in the dark space just a few feet from the ceiling and pulled up the menu on his smartwatch. He set a timer for three hours and let the rattle of cargo and the cars calm him down.

Sam gave thanks that he was able to escape. He knew this journey was just starting. As he faded to sleep, he hoped that he didn't need to jump out in a place where the maglev tracks were elevated. Jumping from a moving train more than twenty feet off the ground wasn't something he figured would end well.

A few hours later, his alarm went off, and he lurched from his sleep with his hand going to his sword. Sam realized where he was and relaxed.

The sound of the wind thrummed at him through the walls of the box car like an old pipe organ used at his church growing up. He decided he would take a peek to make certain he didn't miss his stop.

He climbed back on top of the container by the cargo door he managed to wedge shut and peered out through the slit where the door met up with the side of the train car. He saw no buildings, just trees and other greenery whizzing by.

Sam decided to lay there and wait for a bit. He had to be getting close to his destination.

About fifteen minutes later, the train slowed. Sam peeked out again and saw buildings overgrown with foliage speeding by this time.

He kicked at the container he used to wedge the door shut earlier, and it moved enough for the cargo door to slide again. The time had come for him to jump off the train before it arrived at the station.

Sam slid the door open far enough for him to perch on the side of the container that blocked most of the door and waited. The train slowed further. Sam saw the stationed looming ahead. He needed to leap off the train car as soon as possible in case drones hovered nearby. The train was still moving along at over thirty miles an hour, he guessed. He had no choice.

"This is going to hurt," he mumbled as he leaped off his perch.

He tried to roll when he hit the ground the best he could. His ankle twisted in an awkward position when he touched the ground and went into a roll. He slid through some brush and ricocheted off a tree, and came to a stop.

Sam grunted as he pulled himself up with the help of a nearby sapling. He turned as the train moved on towards the station. Sam winced as a twinge of pain came from his ankle. He leaned over and inspected it, moving it around a bit. Luckily, the ankle didn't break in the fall.

The nanites would repair the sprain within a few minutes; nothing to worry about. He got his bearings and limped off north in the direction of the Ohio River.

The safe house was going to be a four-hour walk after he crossed the river. With the state of his appearance being what it was, he decided waiting until dark would be the best choice. He found an old rusted-out garbage bin tipped on its side.

The rusted bin must have been abandoned years ago. The inside contained bits of refuse and some vermin scat—nothing that would help him. But it would make sufficient cover until dark set in.

He crawled inside and let the nanites finish repairing his injured ankle and other bumps and bruises he received from his jump off the train.

CHAPTER TWELVE

Darkness fell over the area a few hours later. Sam made his way to the Ohio riverbank. The Ohio River did not resemble the Potomac River he had crossed twenty-four hours earlier. This river was a fast-moving tributary of the legendary Mississippi River. Another hundred miles or so downstream, the river doubled in size, and areas of the river contained a rip current that could drag even the strongest swimmer to their deaths.

This location of the river spanned one hundred fifty yards across. The current was still swift. Sam worried he might end up miles downstream if he didn't cross fast enough.

The darkness made locating obstacles in the river difficult. Even with his enhanced ability to see in low light, the river still resembled a black snake with very little details. Submerged hazards posed a serious threat.

Lucky for Sam, the July heat made him yearn for the water to wash away the grit he'd picked up during his leap from the train.

Sam slipped into the river and used his strong legs and

arms to propel himself across. The water remained cool and washed away the buildup of sweat from the humid air. He negotiated the river crossing with no ill effects. Other than his waterlogged boots and clothes, he was uninjured as he clambered onto the bank on the opposite side.

He rested for a few minutes, letting the water drain from his boots. He spent a few more minutes checking his surroundings and located some landmarks to help direct him to the safehouse. Once he had a heading, he started jogging.

After a couple of hours of moving through ravines and getting ripped up by underbrush, he wished for the cool river water again.

When he came upon the location hours later, he realized it was not a house. The safe house was a mobile home. The dwelling seemed an unlikely safe house considering a tree grew through the side of it. Yet, the coordinates matched the safe house location.

Ivy covered the top of the mobile home like a thick carpet. He stayed concealed for twenty minutes, scoping out the area.

Once he convinced himself no ambush awaited him at the location, he made his way to the back door and pried the vine covered door open.

The location had not been visited in quite some time. The ivy had grown over the door and had not been disturbed until now. The disturbed ivy would alert others of his presence. However, he didn't intend on staying long and would be gone by dawn even if he could not contact anyone.

Nothing was clean in this old trailer. Insects of all kinds inhabited the place. The inside of the dwelling resembled a little planet unto its own, with spiders at the top of the food chain feeding on the hapless insects finding themselves caught in a sticky web.

The tip-off for any Reboot safehouse was finding an intact mirror. After dodging spiders and bugs, he found a floor-to-ceiling mirror in what he supposed would be considered the master bedroom.

He sliced a huge spider perched near the mirror in two with his sword. He did not care for spiders, and that spider was almost the size of his hand. He shivered as he kicked the still-twitching pieces away from his feet.

He licked his thumb and pressed it to the lower-left corner of the mirror. After few seconds, a faint beep and click emanated from the wall. The glass lit up, and words scrolled at the top that read:

Welcome Sam, push to enter.

Sam pushed on the mirror, and it sank back into the wall a few inches. The dirty mirror slid to the left into the wall, and he stepped inside a small thirteen-foot wide by seven-foot-deep room. Some small led lamps lit up the windowless room. He slid the door back behind him and investigated the hidden location.

Sam had never been in a saferoom. He had been informed the room would contain everything a field agent might need in an emergency.

Boy did it ever. He stared at a wall of weapons. Projectile weapons of all types encased behind security glass lay before him.

A small section contained some melee weapons. No molecular swords were present. The standard melee weapons would do in a pinch if an agent didn't have one. He did, however, find some rounds for his ST-115.

He poked around a bit more. The room was dusty, but it

had been well constructed. Not a single bug had wriggled their way into the room. A small cupboard contained some ready-to-drink meals such as the military used. A pallet of bottled water was stashed against a wall. Sam grabbed a bottle and chugged the whole thing down as fast as he dared. The water tasted stale but made him feel rejuvenated. He located a few empty backpacks and filled one with food and water.

The nanites kept his body working at peak efficiency. They would maintain an eight percent body fat level and use four percent of the fat's stored energy to keep him functioning. Once he fell below four percent, they would start triggering his body to slow his processes down.

In essence, the nanites put him into a hibernation condition where he still functioned but at a reduced state. It would worsen as he ran low on fat reserves. At around two percent, his body would have no choice but to start eating into his muscle tissue. However, this process allowed him extended periods of time where he didn't need to eat. Water, however, remained essential.

In the back of one of the storage drawers, Sam found a couple of satellite terminals. He put one of the small devices into secure mode and pulled up the news.

The situation was as he feared. All the news stations reported the treasonous acts of Reboot. His picture came up more than once.

The news indicated he killed quite a few people in his escape. That made him a priority target for the local authorities as well as any national alphabet agencies. *Smart tactic by a deceptive enemy,* he thought.

Sam couldn't believe this was happening. Everything seemed to be collapsing around him. He hoped his foster parents still lived and escaped incarceration. The news did not mention much about the employees. No live images were being shown of any locations.

He sat for a few minutes and rested his eyes. His mind wandered from his parents to Alyssa. He hoped she remained safe as well. Dwelling on them was not helping his situation. He needed to figure out what to do next. He had to try to contact anyone left from Reboot.

He decided to give the emergency line a try. He fired up the program on the secure terminal. The terminal asked him to voice match, enter his ID code, and submit multiple biometric scans of his fingers and retinas.

The screen went to black with a cursor in the upper left of the screen that read "connecting." Then it changed to "please wait." A few minutes later, the screen came up to an unlit room. A woman's voice he didn't recognize came through the speaker.

"Hello?" she said.

Sam hesitated. The figure sat in a dark room with her face obscured.

"Sam?"

"Who are you?" he asked.

"My name is Laura."

Sam's eyes narrowed. "Your full name?"

"Laura Cinder. I was a board member."

"I'm going to need a bit more evidence than that after what I've seen on the news."

She sat silent for a moment.

"OK, I'm out."

Sam reached for the button to turn off the terminal.

"Wait! Please, wait ... would your full name be enough?" she asked, almost in a panic.

He thought about it for a moment. Other than his

parents, only someone high up the chain would know his entire name.

"I will accept that as confirmation, Laura."

"Your given name is Samson Elijah Jazeel. Your birth name is Alex Tigwell. When you were a little over a year old, your parents were murdered by a kill squad. You came to us, and the rest, they say, is history."

"OK, Laura. What do you mean you *were* a board member?"

"Up until a couple days ago, I was a silent board member, one of the twenty that remained unknown to everyone in the company except for the other nineteen. I'm forty-eight years old. I became a board member on my twenty-fifth birthday. "I have much to tell you, but I understand time is short, and you need to keep moving.

"The safehouse you're in is known to those hunting you —the government has access to those records now. Your entry into the room would have notified someone of activity at that location."

"I don't intend on staying long. How is the company doing?"

"Priority one was to keep them away from our technology, so we had already relocated most of Reboot's technology by the time they made a move on us. Enough about that for now. This entire situation was planned for. I am aware your memory is photographic. Here is a map of safehouses never included on other lists."

A popup window opened full screen, showing fifteen safehouse locations from Maine through Pennsylvania and down as far south as North Carolina. Each location was listed along with the address.

"Let me know when you've committed the locations to memory."

After a few minutes passed, Sam confirmed he had what he needed.

"How did you know I have a photographic memory? My parents kept that a secret," he said.

"Yes, but they also worked for us. We are the ones who made them your foster parents. In return, they gave us feedback on your strengths and weaknesses. Your intelligence and memory are some of the reasons why you're our best operative.

"Physical strength can be gained in many ways. Mental aptitude, street smarts, confidence, and intelligence all stem from your mind ... genetic limitations in the mind cannot be overcome so easily, even in this age. You need to accept there are certain things you were not made aware of during your youth. I promise you this: once you relocate to one of the new safehouses I showed you, we can spend as much time as you like filling you in everything we understand about the situation, the status of Reboot, the whereabouts of your foster parents, why I believe you're our best chance of success, and where we go from here."

"Best chance of success? Isn't the coup over?" he asked.

"We've not even begun, Sam. As I said, we created a contingency plan ... you. More of that later. Pick a new safe house and make your way there. I do not need or want to know which one you pick. Make it to one and make contact at the following IP."

She held up a piece of paper with a long IP number scrawled on it.

"Are you good?" she asked.

"Yes. I'll contact you once I'm on location. The journey will take some time. I'm going on foot and staying to the uninhabited areas."

"Be careful, Sam."

"You do the same. I need you as much as you think you need me."

He turned off the terminal and slipped the device inside his backpack. He stuffed the bag with as much food and water as he could and packed additional rations into every pocket he could find.

His plan was to jog cross country. If he averaged five miles an hour and kept that pace for eight hours a day, he might cover around forty miles a day.

His best bet would be to move during the day and hide at night when drones had an easier time detecting his heat signature. He located an anti-IR poncho that he could sleep under at night.

It would be difficult to keep that pace with the weight he carried when he started out, but his pace would improve as he ate through his rations and his load lightened. Sam gorged himself on water and food before leaving the safehouse.

The specialized nanites in his body allowed him to do things that would make a normal person collapse. They were not just a cure for diseases like the public vaccinations. His body could maintain abnormal levels of stamina as long as he ingested enough calories to maintain it. The nanites would ensure his cells didn't build up toxic levels of lactic acid and other wastes.

There was always a danger to pushing himself too far, though. Nanites were microscopic computers with a code they followed no matter what. They would not stop to determine what tissue they needed to consume to keep the level of exertion.

He left the safehouse and camped a few miles out until dawn. He gorged himself again on food and water that hadn't fit in his pack and left the rest behind. He carried an additional fifty pounds of supplies that should last him long enough to trek the distance to his next safehouse location.

Before leaving, he pulled up maps of the area between his current location and the target safe house. The one he preferred was in Maine, which was well outside his range. Even averaging forty miles a day, it would take him weeks to make it that far. He didn't have the time or resources for a trek that long.

His target would be the state of New York, some distance south of Buffalo near a pre-war town called Ashford. The town bordered some old national forests that remained devoid of homes. This allowed him to travel unabated.

His route would take him near two other safehouses in case an emergency arose. Ashford was his primary target. Even at that distance, the journey would take him over a week on foot.

Sam headed out as dawn peeked over the horizon. He stayed within a mile of the Ohio River and followed the shoreline north. The journey was long, so Sam focused on the trek like it was a training program back in the academy. He busied himself with thoughts of loved ones, Jake's betrayal, and Alyssa.

He wouldn't be able to go out in public with the entire government trying to hunt him down. At some point, he would be required to change his appearance. He hoped Laura would give him some ideas on that. She did say they needed him, and he was the key. What key she spoke, of he didn't know.

Reboot had some plan in action. Otherwise, they would not need his assistance. He wasn't surprised Reboot had planned for this kind of problem. The company hired hundreds of genius-level people similar to him, and the company was founded on figuring out problems that seemed unsolvable.

The days trudged by, and he covered enough distance north that he could cross the much smaller Ohio River with little struggle and head northeast.

Near the border to New York, he hid from a pack of drones that were headed north toward Buffalo. He stayed under his poncho and the heavy canopies of the forest for cover. A fast-moving drone couldn't tell a tree from a rock in the daylight hours. He was confident they wouldn't be able to spot him, as long as he was careful.

The drones packed excellent thermal systems they put to good use during the night. He had relied on them when he moved Jake during hours of darkness.

The days went by as he closed in on his objective. His long journey had taken him over a hundred miles. His tactical boots, designed for the city, did not take this type of journey well.

The seams were splitting here and there. He was filthy, smelled horrible, and his hair had dirt, twigs, mud, and Lord knew what else in it from burying himself under leaves and dirt at night to hide his thermal signature.

The briars and undergrowth had shredded his pants, jacket, and pack. He had almost two weeks' worth of beard growth. He might not be able to recognize himself let alone anyone else.

His supply of food neared exhaustion. His water supply was good. He was able to refill most of the bottles he emptied after crossing the Ohio River for the last time.

He'd traveled over three hundred miles in nine days,

wondering if the distance might be a cross country record of some sort. The human perseverance when under stress was remarkable.

Sam made it to the address he'd memorized from the list Laura had shown him. The location was off the beaten path, which was why he'd chosen it over the others.

The old farmstead sat on the edge of what he determined might have been a field at some point. It was nothing but overgrowth and small trees now. But the white barn and old farmhouse still stood against the ravages of time.

Mid-afternoon settled in on the area by the time he stepped on the weathered porch of the farmhouse. He tried picking off the hundreds of small briars clinging to his clothes and hair.

He gave up in frustration and walked up to the door and gave it a push. To his surprise, the door swung open.

He removed his sunglasses, pulled out his sword, and wrestled his ST-115 from its holster and checked out the first and second floors. The second floor looked near ready to collapse. That left the basement.

He hated basements. The issue stemmed from a horror movie he saw while in his youth when the doctors didn't understand the nature of his photographic memory.

He could still play back the entirety of the movie in his mind. The movie always bubbled its way to the surface when he had to go in the basement of an old house.

He came to the basement door and wrenched it open with his weapon and sword ready. The stairs led into the darkness of the basement. No power ran to the old home, the power lines outside collapsed long ago from age.

He hoped he hadn't picked a location with a flooded basement. The lack of a mildew smell hitting his nostrils instilled a bit of confidence.

Sam flicked on a small light he'd snatched from the safe house. He strapped it to his head and proceeded down the stairs one step at a time. The wooden steps creaked as he moved down them.

The noise did not make him feel any better. Images of the movie flashed in his mind. He pushed them away and reminded himself he remained a trained operative, and movies were figments of someone else's demented imagination. Of course, that didn't mean he was going to let down his guard.

The steps led down much farther than a normal basement would have been dug. He shined his light on the wall to his right alongside the stairs. He realized new work had been done at this location, indicating he remained on the right path. The air cooled as the stairs came to an end, and Sam found himself in a small eight-foot by twelve-foot cellar.

At the far end, away from the stairs, he found an odd door. He walked up to the massive, sealed door and investigated it. The door was made of stainless steel or a similar alloy. It resembled a bank vault door from the late twenty-first century before paper currency went out of circulation.

A thick handle around three feet high stuck out from the surface of it. No other latch or mechanism for opening the door could be found. So, he gave the handle a tug. The door moved enough to let him realize it was unlocked and quite heavy. He put his weapon and sword away and grabbed the handle with both hands.

He pulled hard, and the door gave a few inches before falling back shut with a thud. He braced his foot against the wall and pulled again. This time he put some leverage behind it, and the door pulled open a few inches. A gush of air rushed through the crack in the door with such force it caused the basement door at the top of the stairs to slam shut.

However, the massive door remained open long enough for him to jam his shoulder against the wall for further leverage and brace for a hard push. The house creaked with the sudden disturbance of air.

The safehouse must have been vacuum-sealed. The vacuum eliminated the threat of fires or anything growing since oxygen was required for most life. He didn't need to worry about the boogieman down here. It also meant nobody waited for him inside.

After wedging the door open, he went back to the top of the stairs and tried opening the basement door. The old wooden door wouldn't budge. He pulled his sword and cut an opening in the door. More air rushed in. That would have to do until the room or rooms below equalized with the same air pressure as the floor above.

He sheathed his sword and walked back down the stairs, peering into the opening for the other door. Perforated metal stairs led down. He waited for a few minutes and went down the stairs. After entering the staircase, an array of LED lights flashed on, half blinding him. They lined the stairs all the way down to the bottom some hundred feet down. He blinked away tears and shielded his eyes until they adjusted to the sudden light.

"Wow," he whispered.

Sam clicked off the light strapped to his head and removed it. He went back to the steel door and kicked out the wedge he'd placed there. It had kept the door open wide just enough for him to squeeze by. Behind him, the door closed with a light thud.

Sam put on his sunglasses to keep the bright lights out of his sensitive eyes. He made it to the bottom of metal stairs, and a much larger room welcomed him. The room had a solid table and chairs, two soft couches, a simple kitchen area, and a media center that all looked as if they had been installed the day before.

A metal alloy wall faced the stairs he'd come down. The room was at least thirty feet square. An outline for what might be an opening was present in the strange wall. No handle or any other method of opening the entrance presented itself.

The lights weren't the only thing automated. The sound of air moving through a couple of vents indicated his motion triggered an automated ventilation system. He walked around investigating the room until he was satisfied with the layout.

First priority was to eat and drink. No running water could be found, but the storage cooler was packed with water and more long-term meals more about sustenance than taste.

Once he finished eating, he moved to the media area. He pulled the terminal from his pack and plugged it in the media center to charge. The tablet powered up, and he sent a message to Laura via the IP address she gave him, indicating he had made it to a safehouse.

Minutes later, the media center powered up on its own. It displayed the words "connecting on secure channel" for some seconds, and then a woman in her late forties to early fifties appeared on the screen.

"Sam, is that you?"

Sam chuckled. "Do I look that bad?"

"Well, you don't look like you're about to go to a state dinner," she said. "Oh, I'm sorry, you haven't seen me before. I'm Laura Cinder."

"Your voice gave you away. Is it safe for you to display your face on a transmission?"

"Yes, this channel is not being transmitted over the air waves. This is a hard connection from your safehouse to my location."

"What is this place? It isn't anything like the other safehouse. This one has purpose. Are all these like this one or am I just that lucky?"

"Yes, all the dark locations are connected to a central hub that manages and monitors about fifty locations. We knew someone found the site when the pressure alarm went off. I assumed the alarm was triggered by you and that you would take some time getting settled and then make contact. I monitored you on camera for some time—I hope that's alright. The system is automated, so I prayed you didn't do anything like decide to clean up from your trip before contacting me."

She giggled at her joke.

"Well, at least you have a sense of humor about the situation."

She returned to being serious.

"Now that you're secure, what I need is for you to meet with me face-to-face for a debriefing."

"How do you figure that happening? The journey took me over a week to make. Unless you happen to be on the other side of the steel wall, I doubt the trip will be quick."

"Well, would you be surprised if I said I was-so to speak?" She gave a wry smile.

"I'm not certain I follow."

A second later, a noise came from the wall. First, the wall clicked and let out a sound of rushing air, and that was followed by a metal-on-metal noise as the wall shifted and sunk in a few inches before pivoting down into the floor.

Sam watched as a man-sized tube slid in on a set of maglev rails and came to a halt.

"Grab your things and climb in the tube."

Sam's eyes went wide in surprise.

"You're serious?"

"Absolutely. This will take you to my location via a hidden subterranean system we installed years before our coup was executed. The rail system is a miniaturized version of the same technology used in the Atlantic run."

"This thing uses the same tech as the trains making the supersonic runs along the ocean floor? Am I going to go that fast?"

She grinned. "Not quite, but fast enough."

Sam hesitated for a moment and then shrugged.

"What the hell."

He grabbed his things and put them in the base of the tube and crawled in. He slid the door of the capsule down, and the faceplate latched into place. Sam watched as the media screen powered off along with the lights. The heavy steel door slid back up in place, and he found himself in complete darkness. He heard the expulsion of air as the tunnel went into vacuum.

Sam quickly realized how ripe he smelled now that he had stuffed himself into a small man-sized tube.

The capsule started moving him feet-first towards whatever destination Laura had programmed into the system. There was no turning back now. At his feet, a screen lit up, displaying MPH. The yellow display gave off a dull glow inside the capsule.

Sam let out a sigh. He hoped the ride wasn't a long one. Not a lot of room remained for him to scratch an itch, let alone sit up. Plus, his body odor made him wrinkle his nose in disgust.

The MPH gauge moved upward at a slow pace at first. Once it hit 100 MPH, the man-sized tube surged forward. The G-forces pushed the blood into his head until he thought we would pass out. The indicator blasted past 500 MPH in fifteen seconds. It hit Mach 1 before the tube decelerated.

The capsule came to a halt, and the MPH indicator shut off, leaving him in complete darkness again. After a few seconds, Sam heard a rush of air and the same metal-on-metal sound of the wall sliding out of the way. Light streamed in, and Laura stood there, waiting with two attendants.

One came over and unlatched the capsule door for him so that he could slide out. He grabbed his possessions and handed them to the attendant, who reluctantly grabbed the items with her fingertips and held them out almost at arm's length in an effort to keep the smell away.

Laura stood almost as tall as Sam. They might be underground and on the run, but she still dressed in a business pant suit and heels. Her slim figure gave away the time she spent exercising. Her fiery red hair was tied into a bun with a silver stake through it to keep the hair in place.

Laura turned to the attendant and pointed to the half-shredded bag.

"He won't need those any longer."

The attendant nodded and quickly took the bag away.

The second attendant came up to him.

"It is a pleasure to meet you, Sam. My name is Robby. If you'll follow me, I will take you to your quarters so that you can clean up and put on some fresh clothes."

Sam looked at Laura.

"It's OK, Sam. Robby will take care of you. When you are finished, he'll bring you to me for debriefing. We need to hear everything you found out about the coup, how the government discovered it, everything ..."

"Yes, ma'am," he said.

Sam followed Robby down a long thin corridor and turned a corner leading to an intersection. Doors lined the hallway for hundreds of feet in either direction.

Robby showed him to one a few doors down the left side corridor. He opened the door for Sam and motioned to him.

"This is your room, sir. You should find everything you need including a shaving kit."

"Are you saying I need to shave?"

"Well, I wasn't going to say anything, but it looks like a bird nested on your face."

Sam gave out a hearty laugh. Turning, he caught his reflection in a small mirror and groaned.

"Shit, you aren't kidding!"

Sam gave the room a quick glance and turned back to Robby.

"I'll be back in forty-five minutes to retrieve you for your debriefing with Miss Cinder."

Sam nodded. "Understood."

Sam closed the door and looked again at his surroundings. The room was twelve feet by twelve feet, which made sense based on the staggering of the doors in the hallway.

Fresh linens sat on a single-person cot in the corner. Opposite the bed sat a shower, sink, and toilet. A kitchenette was to the left of the door and to the right of the bed. The compact space contained everything a person needed.

Even an Echo-Lite razor sat on the sink. Sam picked it up and smiled. It was going to be nice to get that bird's nest off his face. The laser-powered razor was much better than a sharp metal blade he heard people used prior to the war.

He would have shaved his head before leaving the last

safehouse, but long hair and a beard made him harder to recognize. Sam made quick work of the tangled mess of beard with the laser blade.

That left the thousand or so burs stuck in his hair he would never get out. He grabbed a small trimmer and gave himself a buzz cut to free his head of the torturous little seeds. He ignored the few passengers he saw crawling throughout the hair he cut from his head.

He found a small chest under his cot that contained a set of one-size-fits-all clothes. The waistband of the military cargo pants was made of elastic material with a draw string. Sam shook his head. Some things never changed.

The undershirt was a combination of materials made to expand to fit him yet remain comfortable. The pants were a bit long, though. Sam eyed the length and used his sword to trim the legs to something more to his liking.

He shed his nasty clothes and stuffed them into the trash container in the kitchenette. He still had about ten minutes before he expected Robby to knock on his door, so he made his bed, sat down on it, and discovered a small tablet sitting on a fold-out table next to the cot. He picked up the tablet and prepared a quick outline as a reference he would use for the debriefing.

Not long after he finished his outline, Sam heard a knock at the door. His instinct was to grab his sword. He remembered he remained in a safe location for the moment and opened the door. Robby stood in the hallway at parade rest.

"If you will follow me, I'll take you to Miss Cinder."

Sam fetched his tablet and strapped his sword to his waist.

"You won't need your weapon. It's just a debriefing.

The weapon will be safe here," Robby said as he pointed toward the sword Sam had finished strapping on.

"Where I go, this goes."

Robby shrugged. "Suit yourself."

He led Sam back toward the maglev shuttle system and took a left. Four hallways led away from the area like fingers of a hand. Robby paused.

"This is the maglev pod station you arrived in. They run to all our top-secret safehouses. We can shut down the station for every safehouse, and we monitor all safehouses for activity. This prevents any misuse or infiltration by government operatives."

Robby motioned towards the way they came.

"That corridor of rooms we came from are the living quarters. We can house one hundred people at this location or double if we put two to a room. We are stocked with enough food to feed that many people for five years, and we can process an unlimited amount of fresh water. We have a staff of twenty, and the one recent addition puts us at twenty-one for the time being."

Robby motioned to the hallway to the far right.

"That hallway leads to the cafeteria. It's self-serve. "The hallway to the left of that one leads to a training area. Free weights, machines, cardio equipment, and a trauma area for injuries are all available for use at any time of the day or night.

"That reminds me—the lighting is set to follow the day and night cycle. It'll change hue to match the body's response to light during that time of day. For example, it'll change from orange in the morning, to yellow during the day, red in late afternoon, and a cool blue after dark. A digital clock on the wall to our right indicates the time of day at our present location.

"The third hallway over is where we'll go to meet with Miss Cinder. She is waiting for us in a conference room we've been using as a strategy planning area for the past couple of weeks. "The final hallway is where we house our communications station for all incoming and outgoing traffic to other stations and operatives. We also house our intelligence section there that monitors all other stations and intelligence from outside sources. Questions?"

Sam looked around, taking in everything. The facility felt very militaristic in nature. He liked the feeling.

"Thank you for the tour, Robby. I'm ready to see Laura now."

Robby led him down the third corridor from the right, and they passed a couple of smaller rooms before coming to a door across from the main conference room. The main conference room looked large enough to house the maximum occupancy of the station with ease.

Robby knocked on the door.

"Come in," said a female voice.

Robby opened the door and motioned for Sam to enter.

"I'll be in communications gathering some items if you need me, Miss Cinder."

"That will be fine, Robby, thank you."

She looked over Sam with a satisfied glance and smiled.

"Please sit." She motioned to the chair in front of the desk.

"This is an informal debrief. I will give a formal one to the rest of the surviving board members."

Sam looked up at her as he took his seat. "Surviving board members?"

"A lot of things happened since you went rogue and attempted to kill congressman Tillman."

"Ahh ... Is that the spin they're putting on it?"

"Oh, that is just the start of their propaganda, my dear. We knew something went awry when operatives went missing and politicians on the payroll failed to check in. However, the news story stating an attempt on Jake Tillman's life by his own head of security presented the largest argument that the coup needed to be postponed. At that point, we executed Plan Zeta."

Sam nodded to let Laura know he followed her explanation.

"Plan Zeta is a contingency plan we put in place at the same time we implemented the plan for the coup. If the plan were to be discovered by the government, Reboot would be guilty of treason, and the company dissolved, and all assets confiscated by the government. Due to the size and wealth of reboot, we are not in danger of being dissolved. We are a world organization such as NATO and WHO. Our headquarters might be in the United States, but there is no head to cut off."

Sam interrupted. "What happened to our stateside assets after I went on the run?"

"Hours before the newscast, the government moved in on our facilities all over the U.S. with military precision. Unfortunately for them, we were well on our way with Plan Zeta. Our most delicate technology data had been stored in a secret underground facility for many months prior. We purged all data and destroyed as much as we could that remained in hard copy formats. We estimate that eighty-five to ninety percent of all our technical data remains out of the hands of the government. Most of the apprehended data is represented by PPF technology and some molecular sword tech. No nanite tech remained stateside because of the government's growing aggression towards Reboot due to us being unwilling to allow the nanites to be weaponized."

"What about my parents and the rest of the team?"

Laura's gaze instantly grew somber. She walked over and stood in front of him.

"We lost Tony. He was escorting some execs when the government came in and started shooting. He fought back and was killed. Zack is out of country. I'm not certain where. Africa, I think. We didn't expect the military to place a shoot on sight order on our resisting staff members. Many were killed where they stood when the military swarmed in. I'm sorry to tell you your foster parents were among them."

Sam sat for a few moments in silence. He felt his face go flush with blood. A bead of sweat popped from his temple and mixed with the sweat appearing from the rest of his face. He clenched and unclenched his hands. He gazed up into Laura's eyes with tears forming in his.

He whispered, "Those fuckers..."

Laura knelt and put her hand on his.

"I am so sorry. I thought it best if I told you. Your parents and I were close friends. I ... I was present when they adopted you."

Her hand started shaking on his. He put his other hand on hers, and for a moment, they both wept silently together for the loss of friends and family.

After some minutes, they both stood, and Laura got out some tissue, sharing it with Sam. They wiped the tears from their eyes and blew their noses.

Sam walked over and leaned against the wall, and looked up at the ceiling. He gave a few quiet words to God to welcome his parents into his kingdom and let them know he loved them. At the end, he looked back at Laura.

"Anything else?"

"There isn't much else to tell. All our stateside facilities are shut down. Those who escaped from the authorities are on the run like you or went underground like I did in locations such as this one. What we need to discern is what you know."

"Have you heard from anyone else from my graduating class?"

"Not yet, but that doesn't mean anything. We are still getting reports of staff and operatives making it to safehouses and black sites all over the country. The government doesn't have the resources to hit every location at once. That gives us time to secure personnel and regroup."

She paused to hear if he had another question.

After a moment, she asked, "What can you tell me?"

Sam leaned forward and stared at Laura with his eyebrows furrowed.

"I know who is responsible."

"Is it Jake?"

Sam shook his head up and then down one time.

Her jaw set, as if confirming her fear. "Why would he betray us? Do you know of a reason why he would?"

"He gave me some line about how he couldn't take the pressure any longer. How hard it was for him to be two people at once. I'm not certain I believe the story. What I do know is he's responsible for every life the government has taken because of his confession to the congressional members. When I find him, I will snuff out his life like the government did my parents."

"We need to complete a lot of work before you are going to have a chance to do that."

"What do you mean?"

Laura started pacing the length of the room. "Our goal is still the same. The government needs to be overthrown."

"How can we achieve success in the state we're in now?"

"As I said earlier, Reboot is not located in just the United States. It is a worldwide organization with connections with many other governments, the WHO, and what's left of NATO. We established research facilities all over the world. For now, the fight is here, and here we will stay until the fight is won."

"So, we are going to turn into a rebellious guerilla force now?"

She gave a mirthless laugh and stopped her pacing. "Nooo, nothing quite like that. The next step is to get you back in the fight. Doing that will require a special nanite injection."

Sam looked at her. "I'll do whatever is necessary to finish this. I'm in one hundred percent." Sam held up a finger. "But I won't be kept in the dark any longer. I want to know all the details. I'm not going to be a weapon for Reboot without you taking some risks."

She smiled back at him. "I wouldn't have it any other way."

"What now?"

"I think you need to rest. You've had a tumultuous past couple of weeks. I'm going to arrange the necessary things. Tomorrow you will go into surgery, and when you wake up, you will look like a new man."

"Interesting ... A new man, you say?"

Sam walked to the door, stopped, and turned back and walked to Laura. He reached out and gave her a hug.

"Thank you," he said.

She hugged him back. They pulled apart, and she smiled at him.

"We'll talk more tomorrow. Get some rest, kiddo."

He went back to his room, showered and crawled into the bed he had been longing for since arriving. He drifted to sleep while his mind wandered over Jake, Reboot, Laura, his parents, and the fight to come.

He woke the next morning to a light knock on his door. He expected it to be Robby, so he opened the door wearing nothing but his boxer briefs. To his surprise, Laura stood there.

"Whoops, my apologies. I thought you were Robby," he said as he grabbed some pants.

"Don't bother putting those on."

He looked back at her with a questioning glance.

"Put this on instead." She tossed him a hospital gown.

He slipped the thin gown on and grabbed a fluffy white cotton robe hanging next to the shower and slipped it over the hospital gown while she waited in the hallway. He stepped out on the cold tile floor. She gave him a once over.

"You'll have the ladies swooning for you in that," she teased.

"Shut it," he said in jest.

They started walking down the hallway towards the infirmary where his surgery would take place.

"Tell me about this procedure."

"This nanite injection will have one hundred times as many nanites in it than the ones in your body now. The difference is that they're engineered to reshape bone, specifically the ones in your face."

"That sounds unpleasant," he replied with a slight frown.

"Oh, the pain will be excruciating. That's why we'll be putting you into a chemical-induced coma for the duration of this. "The surgery is going to take days and involve a lot of pain. The nanites increasing healing will work in tandem with the nanites working on your facial recon-struction.

"Once the procedure is done, the specialized nanites will deactivate, and the body will remove them like it does any other type of waste."

"Has anyone had this procedure before?" asked Sam.

She smiled. "One other. He's the reason you are getting put into a coma."

"Ah. Sometimes it sucks to be the one pushing the boundaries of technology."

"You said it."

They entered the room where the surgical procedure would take place. A bunch of medical machines half-circled an operating table.

"When will the doctor be in?" he asked as he took off his robe and climbed up on the table.

"You're looking at her."

She walked over, washed up, and slipped into a surgery suit and mask.

"It seems funny we still dress as if the sanitizing methods we use mean anything at all. I guess old habits die hard."

A couple of attendants assisted her with getting an IV into Sam and laid him back on the table for the injection. While they tended to Sam, Laura prepared the injection of nanites.

Lying back on the table, Sam looked over at Laura. "How long did you know my parents?"

"I knew them before they adopted you. I was the one who hired them. In a way, I got them together."

She paused for a moment, and Sam could tell she was smiling behind her mask, lost in memory.

"They arrived to be interviewed and met in the waiting

area. I ran late from a meeting, and the two of them began chatting. Once they were hired, one thing led to another ..."

"I remember them telling me that story. They left out the part of them meeting you."

"Well, they sort of had to. You know how secretive we are. I consider them my closest friends, and they never realized I was a board member. At times, it felt odd keeping that secret, especially when they made me your godmother."

"Wait, you're the one they always spoke about?"

Her eyes flicked to his. "They mentioned me?"

"They spoke of my godmother on many occasions, but I never got to see a picture or know who you were."

"I suppose it was best for the company. Still, it feels good that they did hint at my existence. I hesitated even telling you in fear that you'd think I was trying to influence you by telling you some yarn about me being your godmother."

"I'd like to talk more about this when we get the chance," Sam replied, watching the assistants prepare his surgical pack.

"That would be something I would like as well."

She walked over with the huge syringe. "I suppose we should get on with it. You ready? This is going to hurt."

He clenched his hands in anticipation. "Go for it."

She slid the needle into his neck and pushed the plunger down.

"How is that? You feel OK?"

He exhaled. "Not bad for having a golf ball injected into my neck."

She leaned in a bit toward him and whispered, "In my office, what did you pray for?"

He paused for a moment, a bit thrown off by the sudden question.

"Well, I prayed my parents to be welcomed into the kingdom of Heaven and that I loved them."

"It is good to see you are still religious while being a scientist."

He shifted his head to look at her.

"I discovered long ago that science and religion are linked. Only a being of infinite intelligence could have created the cosmos and all the things we interact with. Too many things exist that cannot be explained without the existence of a creator including us, for God not to exist. Look how long it took for Reboot to invent nanites. Yet the human cell by itself is infinitely more complicated than they are."

Laura chuckled. "Did you know I am the one who named you Samson Elijah Jazeel?"

Sam's eyes widened in surprise.

"Oh, yes. I named you. I had a bit of a fight on my hands to give you that name. Your parents couldn't figure out why, and I couldn't tell them why it had to be that name."

"Why did you name me Samson Elijah Jazeel? It *is* an abnormal name."

"One day I will explain why, but today is not that day. I assume you researched the religious meaning of it?"

"Of course. Samson was a biblical character who picked up the jawbone of a donkey and killed a thousand Philistine soldiers on the battlefield. Apparently, he wasn't a nice man either. Elijah was a man who stared down an entire nation. Jazeel means 'strength of God.'"

"Yes. Your name is a man with the strength to kill thousands while staring down a nation, because he has the strength of God on his side. Ponder that for the time being. You will understand when the time is right."

One of the attendants handed her two syringes.

"Great, more shots in the neck," Sam huffed.

Laura laughed. "No, no. These go into the IV and will induce your coma. Start counting down from thirty. Before you get to zero, you'll be out. When you wake up, you won't be in any pain, and the procedure will be finished."

He started counting as Laura injected the first and then second syringes of liquid into the IV tube running into his arm. He passed out before he counted past twenty-three. Laura looked over at the attendants.

"Apply the helmet and mask. Once you finish, we will activate the nanites."

Laura ran her fingers through his freshly buzzed hair and scanned over his face one last time as any loving mother might to their child. If what she had been informed of was true, he would be the key to humanity's survival.

"I hope he likes the new face I picked out for him."

CHAPTER THIRTEEN

Five days later, Sam woke up in the infirmary area. Laura and Robby were standing in the small room talking when he came around.

"Take it slow, Sam. You've been out for almost a week."

Sam slid his hands over his face while trying to shake the groggy feeling he had. Then it clicked his face had changed, and he jerked them away.

"It's OK, you can touch your face. Talk, smile ... everything you did before you went under. It took a bit more time than we expected due to the density of your bone structure, but everything went well."

Sam hesitantly touched his face as she spoke. He didn't feel much different. His jawline seemed off. He opened and closed his mouth a few times.

"Why does my jaw movement feel different?"

"I redesigned your jaw to prevent blows to it from knocking you out. It will make you more resilient in hand-to-hand combat, should that ever happen to you."

"Yeah, that's never going to happen," he joked.

Sam stood and walked over to a small frameless mirror hanging on the wall and looked at his reflection. His jaw was more pronounced and squared. He looked a bit like a street brawler with a menacing appearance he didn't have before. His baby face was gone, but his eyebrows and eye color remained the same.

"I went with a rugged look," said Laura. "I hope you're satisfied with it because we only had one shot. Your retinas were also slightly altered to fool retinal scanners. We didn't have enough nanites to do anything with your voice. Other than that, what do you think?"

"I'll get used to it. At least I can go out in public and blend in now. That leaves the question as to my new identity."

Laura looked over at Robby, who sort of stared off into space. After a few seconds, he realized Laura and Sam were looking at him.

"Oh! Yes of course."

Robby handed Sam a small yellow envelope.

"This contains your new state ID with your new name, Sammy Berchester. Your passport, birth certificate, and a small pamphlet about your background are included. The rest you can make up on your own. I assume you won't have any difficulty remembering any of the new information."

Sam looked up from the packet.

"Was that supposed to be a joke?"

"Yes, but not a very good one."

Sam glanced over at Laura out of the corner of his eye. She hid a smirk as best she could.

"Anyway, you are a carpenter and inventor based out of upstate New York. There is a house set up at the location you will be living out of, and another safehouse a mile or so down the road you will work out of. An underground tunnel connects the two locations. Any questions? Ok then, I'll go back to shutting up."

Sam gave him a grin.

"You're alright, Robby."

Sam then looked to Laura.

"What's the plan?"

"For now, you need to lay low. Do some training and keep tabs on the news to see if you can figure out what's going on with the Reboot situation. I will be in limited contact with you. You will only contact us from your safe-house. Consider everything else monitored by the government. The address is in the packet, and your vehicle is waiting for you at a drop-off point we designated for you." Said Laura.

"That is all? I'm leaving that quick?"

"We need you off-premises as fast as possible now that your new identity has been established. That leaves no correlation between us and the new you."

Sam nodded. "I see. I'll clean up and be on my way then."

Laura dismissed Robby. She closed the door and walked back to give Sam a hug.

"Be safe. I don't have to say how dangerous it is for you or any Reboot employee out there. No matter what you see on the news, do not do anything motivated by emotion. The time will come for revenge, but now is not it."

"I understand," he said.

"The packet contains our communication times along with other informational items. Other than that, you are free to go."

"Thanks for everything, Laura. I'll do my best to not let you down."

"I know you won't, my dear. I don't think you have that gene."

She smiled and waved goodbye as he walked through the door.

CHAPTER FOURTEEN

S am went back to his room and found some new clothes stacked on his bed. He changed and packed up his few belongings in a duffel bag. Catching his reflection in the mirror over the bathroom sink, he paused to inspect it.

Sam brushed his hand over his face. The new jawline, bridge of his nose, and cheekbones had changed the most. It felt like he was dreaming.

A lot of time would need to pass before he would recognize his own face again. He hoped that he could go back to his original face at some point after all was said and done.

He showered, shaved, and finished packing the last few bits of equipment. Sam examined the room one last time to make certain he didn't leave anything behind and walked out into the hallway pulling the door closed behind him.

He took the tube transport back to the safehouse he came from. The tube felt as cramped on the ride back as it was the first time, though he was grateful his body odor was absent on the ride back.

A thought crossed his mind as the tube accelerated. He

never learned where the base was that Laura worked out of. Best not to, he supposed.

The innovation required to develop this system of tunnels amazed him. Even more stunning, Reboot kept the whole system hidden from the rest of the world.

Once he arrived back at the safehouse, he found the vehicle Laura said would be waiting for him. A 4x4 sat in the drive. How they got it there, he couldn't figure out. His best guess was they flew it in with a cargo drone.

From the looks of it, they snagged an abandoned chassis from a ditch somewhere and updated the truck for serious off-road duties.

A lift kit gave the vehicle an additional four inches of ground clearance. The drive train was replaced with a new magnetic system. The magnetic wheel hub replaced the spring and shock-based systems from pre-war time.

A magnetic rim, which was nothing more than a metal hoop with electromagnets attached to it and a synthetic rubber band for traction, went where the alloy wheel would have been installed.

The axle contained an electromagnet that repelled the outer rim. The new axle made the vehicle's suspension almost bulletproof and allowed large amounts of travel before bottoming out.

The Department of Defense was still running suspension systems like this through trials before they determined if any would be a satisfactory fit or not.

Sam opened the back hatch and tossed in his duffle bag. The interior of the truck was all business. A spray-on coating allowed him to wash the interior out with water instead of worrying about expensive things like leather. The seats were crafted from carbon fiber and covered in Kevlar.

Leaning in further, Sam could see the interior had a

black finish with an integrated roll cage for additional support and strength.

The exterior was painted a two-tone black and dark green. The paint job blended in well with the countryside but would stick out like a sore thumb in an urban environment.

He jumped in the truck and pressed the ignition button. It clicked on with a hum as the dash lit up with gauges and dials.

Sam laughed at seeing all the tech around him. Based on readout, he could tell the truck was powered by a hydrogen cell-based engine. Clearly, Reboot had spared no expense. Sam put it in gear and slammed the accelerator pedal to the floor. The truck spun all four wheels, ripping up sod and throwing clots high into the air. The broken road and harsh terrain presented no challenge to this off-road beast. He grinned ear-to-ear as he tore off into the wilderness.

After familiarizing himself with the vehicle, he typed in the destination of his new home into the navigation computer, and it laid out the best path based on up-to-the-minute information on road conditions, weather, and terrain.

It scanned the area ahead with 3D mapping technology and made proper corrections for the best route through whatever mess he might encounter. Massive cutting jaws mounted to the front of the vehicle could cut anything it could get its teeth around—perfect for ground debris blocking the path.

Laura had outfitted him with everything he might need for his next task, whatever that might be. Sam welcomed the technology. The more he had at his disposal, the easier his mission would be.

He shook his head and thought about that for a moment. He didn't want to depend on technology. He decided right then that he would force himself to train without the tech in case he found himself in a position where he didn't have it, like when he'd escaped from the capital.

Sam put the truck on autopilot and set about reading the information in the packet Robby and Laura provided him. The packet contained maps of the surrounding area of both his house and the safehouse down the road.

His background story listed him as an orphan again. His parents were killed in the pandemic after the war. He received a nanite vaccine at the last possible moment, saving him from a resistant strain of the flu.

Sam then grew up in tough orphanages during the post-pandemic era that were understaffed and underfunded. There were very few records from that time, which prevented deep investigations from proving his background false.

They had him running away at sixteen and becoming a bouncer. Later, he invented a couple of melee-style weapons that became popular on the black market and underground fight scene.

Sammy Berchester decided to leave that life and return to nature to live in peace. His public bank account statement showed he accumulated a little over three million dollars in cash assets.

The money would be sufficient to explain the truck, house, and the property. That kind of money didn't allow a person to live in luxury. It did, however, give a person peace of mind.

He finished committing the material to memory and switched back to manual drive to continue familiarizing himself with his new vehicle. The truck was amazing. It absorbed terrain no vehicle with normal spring and shock suspension ever could.

Not far from his destination, he turned onto a country road made of gravel and grass that resembled an old logging road. Sam stayed on the old road for a couple of miles until he came to his new home.

A geodesic dome house with dark green aluminum shingles and brown brick stood in a wooded area thick with trees.

A grass drive went up to the house. Everything was a bit overgrown. He would need to take some time to whip the place into shape. That should help him pass the time while he waited on word from Laura.

He pulled up to the garage and parked the truck. Sam walked up to the front door and grasped the handle. After a second, a click could be heard from the lock. Sam pushed the door open and walked in to inspect his new surroundings.

The dome appeared to be around thirty feet in diameter. The house had around nine hundred fifty square feet on-grade and a loft with a spiral staircase leading up to it. A wood stove stood alone near the center of the great room with a chimney going from its flue to the exit at the top of the dome.

Reboot had spruced up the place with furnishings, a complete kitchen, one bedroom, a large bathroom, and a loft. A storage closet and triangular window looked out into the forest from the loft.

Upon closer inspection, Sam realized the house was prepped like a fortress. He could tell that something was different about the windows and the front and back doors.

The walls and dome ceiling approaching two feet thick.

Not much other than an air-to-ground missile looked as though it would make it through.

Upon inspecting the laundry room, he opened a closet, instantly realizing the closet didn't have a floor. A brass pole disappeared down into the darkness below. He slung his sword on his shoulder and slid down the pole. Before making it to the bottom, some lights lit up a room.

The room was storage for long-term foods, weapons, and ammunition. A titanium-reinforced door stood at the other end of the room, away from the pole he'd slid down.

Sam assumed the exit led off to the tunnel Robby mentioned in his briefing. Sam decided he would unpack and relax a bit before investigating the tunnel.

Sam found no stairs leading from or to the underground. To get back up to the ground level required a person to shimmy back up the pole. Getting things up and down would require a bit of effort.

I guess that's what they created rope for, he thought.

CHAPTER FIFTEEN

His first night asleep at the house, he fell into a dream state. The government started turning the United States into a police state. Sam saw visions of riots, military personnel killing civilians, and fires burning unchecked in many cities.

He woke with a start in his bed soaked with sweat. He couldn't believe he recalled the entire dream. Sam never remembered a dream in his entire life. The hair on his neck and arms stood straight out. It gave him a bad feeling.

He walked to the bathroom and splashed some water on his face. Yet, he couldn't fall back asleep. He just kept tossing and turning the dream over in his mind. It felt so real.

CHAPTER SIXTEEN

S am shook off the memory of the dream and did some cleaning around the outside of the house. He parked the truck in the garage and spent the morning cleaning up the briars, weeds, and small saplings in the area surrounding the house. Once he finished outside, he moved back indoors and thought about the underground below the house.

The next morning, he was due to contact Laura at the safehouse. Before then, he figured the room under the house would be a great place for training. He started work on a Wing Chun practice dummy or *muk yan jong*, did some calisthenics, and practiced with his sword.

After he showered, he made his way back to the metal door under the house. Sam grabbed the handle and held it for a moment as he did the front door and the door opened for him.

A dark tunnel lay before him. He found a press plate on the wall and gave it a touch. Lights lining the corners of the ceiling started blinking on away from him and into the distance.

He remembered Laura saying the safehouse lay a mile from this location. The path appeared to be a straight shot from here to the other end. With his enhanced vision, he could see the door on the other end identical to the one he'd just opened.

The tunnel was finished in a fancy cream-ecolored marble tile with dark grey grout. It occurred to Sam that it would be a convenient location to run in private. Plus, he needed to time how long it would take him to sprint from one end to the other. If his cardio remained close to what it was during training, he might manage something near four minutes.

Sam decided he would do that another day. He didn't feel like getting sweaty so soon after showering, so he walked the mile to the other side. The tunnel was large enough to drive a vehicle down—not the type of tunnel he envisioned when Robby told him a tunnel lay between the two locations.

The door at the far end resembled the one under the house. The one difference was no handle presented itself. Just a touchpad existed on the door where the handle would be. He pressed his hand to the pad and waited for the click.

The door opened, and he peered into a small dark room. He pulled a flashlight out and stepped into the small space. The light from the tunnel and the flashlight illuminated the small area that couldn't be any bigger than six foot by six foot square. He found another touch panel on the wall and touched it.

LED lighting popped to life in the cramped room, blinding him for a moment. A braided metal ladder hung from the ceiling on the far side of the room.

He inspected the ladder and found it led to a hatch on

the ceiling. He climbed up and pushed at the hatch. It gave way easy enough. He popped his head into another dark room that lit up with lights upon his movement. Screens on the wall came to life, and the hum of electronics greeted him.

He climbed out of the hatch entry and into the room. No windows appeared in this small room either. Another metal hatch with a braided metal ladder led up on the far side.

The room was shaped like a surveillance vehicle. It was long and narrow with another ladder leading up at the far end.

Three rolling chairs lined a built-in desk mounted straight to the wall under the displays. Before he investigated much further, Laura's face popped up on one of the monitors along the wall.

"I see you are familiarizing yourself with your new home and safe house," Laura said.

Sam leaped back at the sudden sound of her voice.

"Jesus! You scared the shit out of me!"

"No, this is Laura. See ...? Laura," she said while waving her hands around her face.

"Haha, very funny."

Sam relaxed and responded to her initial statement.

"I like what I see so far. The layout will come in handy for training and staying under the radar."

"I admit the safehouse really isn't a house," she replied.

"Come on, Laura, none of them have been a house. More like run-down heaps with a high-tech closet hidden inside."

Laura laughed. "Well ... do you prefer a safe location or an unsafe house?"

"I'm not complaining. Just making a point."

"What I mean is this current location you are sitting in is buried underground. The ladder at the far end leads up in the middle of a grove of small pine trees and bushes that cover the entrance. The safehouse is more like an escape hatch than some place I would expect you to enter and exit on a regular basis."

"It will do fine. I will take this over the rotting trailer I was at when I was on the run," Sam said.

"The house was the primary focus. We didn't want anything too obvious in case someone broke in."

"I hate to be the one to break the news to you, but the front door is a dead give-away that something isn't normal with the house. Any trained operative will pick that out right away. That said, it is still decent cover. If an operative is investigating my front door, my cover is blown anyway. At that point, I'll need a solid door between them and me."

"Point taken. For now, we need you to train up, get used to your new identity, and familiarize yourself with the local area."

"What will you be up to while I'm doing that?" he asked.

"We'll be running down intel on Jake and his handlers. We have an idea for your new identity. It will be dangerous. If you can pull the mission off, you will be in a prime position to execute our next stages of the operation. I'll give you more details on it at a later meeting. For now, do those things I mentioned and maybe try and relax a bit. Get back in touch with nature and establish some one-on-one time with Jesus. It's obvious that it's needed after mistaking me for him," she said, throwing in a wink and a smile.

He ignored her poke. "Sounds like a plan."

"Good. I'll talk to you in a month. See ya, kiddo."

The screen went blank. Sam sat down and leaned back in his chair. He stretched and kicked his legs up on the table. It was time to get back into combat shape. He had lost a lot of weight while on the run. It was time to pack on some muscle.

CHAPTER SEVENTEEN

As Sam continued to train and scout the local area, the dream that occurred the night before he spoke with Laura continued to repeat itself, getting more detailed every night.

On the eighth night, a figure appeared to him in the dream. The person or entity was hard to make out at first. The more times the dream occurred, the better he was able to see the person.

He had a feeling the person might be an angel or some sort of holy person as they seemed to be leading him through the dream without saying anything.

On the fourteenth night, the angel introduced himself.

"Hello, Sam."

"How do you know me?" Sam replied.

"We all know of you. Your name is Samson Elijah Jazeel. Laura Cinder watched over you for us since you were adopted by Reboot. You are on a journey that will affect every life on the planet."

"Who are you ... and what?"

"My name is not important. If you must attach a label to me, call me Paul."

"Do you mean apostle Paul? One of the first followers of Christ?"

"If that makes you feel at ease, then yes, I once went by that name."

Sam shook his head.

"I hope I don't have a brain tumor."

He shrugged his shoulders.

"I'll play along. Why am I dreaming this dream over and over? I don't ever remember having any dreams until now."

"You are remembering this dream because God decided the time has come for you to remember. What you are seeing is part present and part of what is to come if you fail."

"Doesn't God know what will happen?"

"Yes, he knows every scenario of what is to come. God is timeless. Yet, He cannot affect the free will of humanity.

The path has yet to be chosen. All paths lead to the coming of Christ. Yet, every path leads to a different number of those who can be saved.

God loves all his children. But, like all parents, he cannot make them love Him back. He does what He can to save as many as possible before the time for judgment comes."

"What must I do?"

"For now, look upon this dream, this present and potential future to come and steel yourself for what must be sacrificed to prevent this future."

"I will do anything and everything to prevent this from happening."

"Time will reveal if that is the truth to come. You have many choices that will affect the future."

"To give you a better understanding of the pain in the world, I am going to give you a small sample God feels from His children at every moment, so you understand why He loves them so and why he wants you to assist Him in saving as many as possible."

Paul reached out and brushed his hand over Sam's forehead. Upon touching his head, Sam felt the dying, the diseased, screams of torture, pleading, crying, children screaming, and it all came to him in a rush of emotion so strong they knocked him right out of his dream and off his bed onto the floor.

He screamed until his throat hurt. Tears poured from his eyes. Sam never experienced anything like that in his entire life. The loss of his friends and family didn't scratch the surface of what he'd just experienced.

As the feeling subsided, he curled up on the floor next to his bed and wept. He fell back asleep from the exhaustion and woke in the middle of the afternoon.

For the first time in as long as he could remember, his body was sore. Sam clenched so hard during the dream that his body was still recovering. He realized he was out for over sixteen hours. He laid down to sleep around ten the night before, and it was past two in the afternoon.

His first priority was to eat. While he cooked up a big meal of blackened salmon filet, rice, and steamed broccoli, he kept playing back the dream in his mind. Had all that happened? Did he meet and speak with an angel of heaven? Is everything he said true?

Sam finished his meal, showered, shaved, and tinkered with some weapons downstairs. He felt drained. He was afraid to go back to sleep and fought the urge until past three in the morning when he fell asleep at the workbench with a piece of a trigger mechanism in one hand and a rag in the other.

The dream repeated. Government agents and the military were ripping people out of their homes. Some were being shot on site or beaten unconscious. Then, Paul returned.

"Am I to experience this dream for the rest of my life?" Sam asked.

"No, just until you realize how far humanity has fallen." Paul reached out towards Sam.

"Wait!"

"Yes?" Paul asked.

"I don't want to feel all the pain again."

"You must. You need to understand the pain and sadness before you can know how to cure it. This will continue until you accept what the world has come to in your heart and in your mind. Humanity has come to the brink and survived again. However, it will come again soon if no one acts."

"This is how much pain God experiences all the time?"

"No. The emotion you feel is but a fraction. If all of humanity's suffering were to be forced upon you, your mortal mind would not tolerate it, and without the mind, the body dies. You will receive more and more as you figure out how to deal with the emotional surge. Then, we'll move on to what comes next."

"What is that?"

"Hush now and focus," Paul replied as he brushed his hand across Sam's forehead once again.

The surge of pain, suffering, and prayers of those in need hit him harder than the night before. Sam resisted for a few seconds. Then, the scream from the night before rose from deep within him. It was a mournful howl that knocked him out of his dream and backward out of his chair at the workbench and onto the floor. He wept until he passed out.

This continued for over a week. He stopped shaving, showering, and ate very little. Every day he fell asleep, he would dream and meet with the angelic Paul. Every time the experience became more intense. By the end of the week, he started to think he was losing his mind.

By the fourth day, he dreaded falling asleep. By the seventh day, he attempted to stay up for as long as possible before falling asleep. By the eleventh day, thoughts of suicide started to creep into his thoughts just to end the emotional onslaught.

He crept up on two weeks since the dreams started. Three days had passed since he last slept. His eyes were sunken, his face grizzled with facial hair, his hair was unkempt, and he stunk from not showering in over a week.

Sam sat in a chair and stared off into nothing. He sat and tried to fight off the sleep that was working hard to overtake him. His body flinched as he felt himself falling asleep. He tried to get out of the chair. But fatigue took its toll. Instead, he fell to the floor and managed to roll over on his back. Exhaustion overtook him.

As he passed into his dream, he was met by the angelic Paul. Sam started weeping.

"Please ... please don't make me go through this any longer," Sam pleaded.

"I see you are starting to understand the plight in the world."

"God is my savior. This I know, but why am I to repeat this experience every time I sleep?" Sam asked.

"That is not something for me to explain in further detail. I will leave that to Him." Paul motioned towards a park that appeared around him."

He wrapped his arm around Sam and ushered him to a bench sitting in the middle of a glade in the park. Sam felt at peace as they entered the area. They no longer saw people getting murdered in the streets, beaten with clubs, or tortured by government agents.

The sun streamed in through the trees, the birds sang songs, and a lone figure walked into view from the far side of the glade. Sam realized Paul was gone, and he and the other figure walking towards him were alone.

The man who came before him wore a vivid white robe. He wore a beautiful sash that could only be described as having every color upon it gleaming brightly in the sunlight. He wore no shoes and smiled at Sam as he approached him and sat across from him on another bench a few feet away.

"Hello, my son."

Sam hesitated for a moment. "I know this may sound stupid. Are you Jesus?"

The man smiled back at him. "Yes, and you are Samson Elijah Jazeel."

Tears streamed down Sam's face. He bent over, grabbing Jesus's hands in his.

"Why am I being forced to experience all this?"

Jesus gently pulled his chin back up to meet his gaze.

"It is not my intent on causing you pain. I needed you to glimpse the state of what might be humanity's future and parts of what has occurred in your time. Without that knowledge, you would not understand what you are fighting for and why."

"Why me?"

"I chose your spirit before you were born. The paths you have taken to this point have all been yours to choose. Many of those paths all led to this point. You are the greatest mortal weapon I have against the evil spreading through the world. Yet, you cannot hope to succeed on your own. I placed those who you can trust in positions to assist you.

"You are one of the few mortals left on Earth that have met me while still living. Thus, you have inside knowledge and with that, power. You will need every ounce of that power to make this through to the end."

"Did you place Laura Cinder?"

"Yes, she and others have helped guide you and aid you on your path. You are to be my greatest warrior for truth and justice. Your name wasn't given to you by coincidence. It was chosen not as a name but as a definition of who you are to become. Your tragedies and triumphs define you. Yet your biggest tragedies are the losses of both your birth parents and your foster parents."

Jesus waved his hand, and an image of both sets of his parents appeared in front of him. They were all talking, smiling, and laughing together.

"Both sets of your parents stood for something. They had good morals, believed in me, and love you as I do."

Tears of joy streamed down Sam's face. He stared at his parents. They looked as young as he.

"Are they in Heaven?"

"Yes, they are."

"Can I meet them?"

"Where they are, you cannot go. Know they will be waiting, and when it is time, you will be embraced by all your loved ones again."

Sam watched the image fade. He stared at Jesus with a serious expression on his face.

"What must I do?"

Jesus leaned forward and clasped both of Sam's hands in his own. Sam felt the warmth and love envelop him.

"You have many tools at your disposal to assist you: friends, contacts, technology, and money. Yet, that will not be enough. I'm going to bless upon you four gifts. You will need to learn to use them just as you learned to walk and to speak as a child. First, you will be able to pull those in need from the jaws of death by curing their illness or healing their wounds. You will not be able to heal yourself in this manner. You are the conduit through which my healing power flows and, as such, cannot be affected by it.

"Second, you will be able to slow time to the point where you can stop a leaf from falling. "Third, you and the shadows will be able to become one. This will allow you to move undetected by eyesight in shadows. You will still be vulnerable to detection by other means. "Finally, I give you the blessing of strength. The strength will not last, so use the power when it matters most," Jesus said.

"Why can I not use these at will or have more power all the time?"

Jesus smiled. "These are limitations set by the rules placed to create the universe. You are but a man and as such are limited by those same rules. I can allow you to bend them but not break them. Some of them you will never be able to use for long periods of time, and you must rest after each use. Practice will allow you to improve the length of time these skills can be used or find new inventive ways to use them."

"How will I know how to trigger these skills you are giving me?"

"When you clear your mind, the knowledge of how will follow."

Sam hoped for a better answer than that or maybe an instruction manual. He realized he wasn't going to receive either.

"Are you ready, my son?"

"I am ready."

Jesus's grip tightened on Sam's hands. "This will not be pleasant."

Sam felt his hands getting warmer where Jesus clasped them. Heat seemed to be flowing from Jesus into his hands and up to his arms.

At first, the warmth felt pleasant. The pleasant feeling changed to discomfort and then to pain. The heat reached his shoulders and moved down through his body towards his feet. He gritted his teeth as the pain continued to increase. The heat made the palms of his hands feel like they were on fire.

Jesus closed his eyes and had a peaceful expression on his face. The pain became as such that Sam involuntarily tried to wrench his hands away. Yet, he was locked in place as if someone glued him to the bench in front of Jesus.

His teeth grinding turned into grunts of pain. His grunts of pain turned into a faint scream. The scream continued to crescendo until it became a howl.

Sam involuntarily strained with every ounce of strength to break the connection between Jesus' hands and his own. Yet, he was locked in place. The smell of burning flesh permeated the air. His hands sizzled like bacon in a frying pan. Sam clenched his eyes shut and felt the burning pain burst from his feet.

Right at that moment, the burning ceased, and he came slamming down on the floor of his house. All the air was knocked out of him, and for a moment, he couldn't breathe. He gasped for air and rolled over on his side. What felt like an eternity passed before his first breath came, and he passed out moments later.

Sam woke a few hours later. He didn't know what to

make of the dream he had. He could not be certain if he believed it had happened. Sam clambered to his feet and stumbled to the bathroom to rinse his unshaven face.

As he slapped water on his face, he noticed burn marks on his hands. Both hands had burn marks right through the palms of his hands to the other side.

He examined his feet and found his shoes were charred right through. At that point, he knew the dream was real. It happened. He spoke with Jesus, and Jesus tasked him with continuing the mission. Sam clenched his hands and brought them to his lips. A tear ran down his face. He had been chosen to be the tip of the spear.

Sam placed his hands on the edge of the sink and looked at his face long and hard. It was time. He hadn't gotten any instruction other than to clear his mind. His primary task now; to figure out how to use his newfound powers and improve his skill in activating and using them.

The old meditation practices he learned when studying Wing Chun were going to come in handy.

The first step was getting healthy again. Since the dreams started, he lost a lot of weight. He didn't look healthy, and he didn't feel healthy. He took off his charred shoes and tossed them in the trash on his way to the kitchen to prepare a meal.

CHAPTER EIGHTEEN

S am spent the next couple of weeks getting back into shape. He had a debrief setup with Laura at the end of that time, and he didn't want to alarm her as to the physical and mental state he had been in for the past few weeks.

He ran sprints in the tunnel between the main house and the safe house. He bulked back up by eating six meals a day and weight training along with training in Wing Chun every day.

On the day of the debrief, he made his way down the mile-long tunnel to the safe house and made himself comfortable in one of the rolling chairs while waiting for the time to arrive.

He made the secure connection at the predetermined time, and Laura's image popped up on the screen. She gave him the same smile she gave every time they spoke.

"It's good to see you," Sam said.

"You as well," Laura replied "You're looking much better. You had me worried for a bit."

Sam tilted his head at the response.

"Are you watching me?"

"Covert cameras are installed at every location for security purposes."

Sam shifted in his seat.

"So ... you saw what happened to me a couple of weeks ago?"

"If you mean this ..." Another screen came up to the right of the one she was on, and the video showed Sam lying on his back on the floor. His bearded face and unkempt hair made him look like a hobo. What happened next was nothing short of amazing.

Sam's arms were down at his sides and what seemed like some hidden force outstretched them perpendicular to his body, and his legs came together. After a few moments, he started shaking, and his body levitated up off the floor about three feet and hovered in the air.

Steam started coming out of the palms of his hands and then beams of light. His shoes started smoking until light burst out of them. Sam's mouth was agape in a silent scream, but his eyes remained closed. All of a sudden, the light went out, and he came crashing down onto the floor. He lay there smoldering for a moment before he started moving.

Laura paused the video. "Yes, I happened to notice that. Care to explain?"

Laura's face remained emotionless.

"I am having a hard time making sense of it myself. I— This is going to sound crazy ... I started having dreams not long after arriving here. The dream was the same every night. I kept seeing what looked like the future. At the same time, I saw things I know have happened: people getting murdered, military taking over Reboot locations, riots, police actions ... just crazy, crazy stuff."

Laura just stared at him, so he continued.

"After a time, I started seeing what ended up being and the apostle Paul from Heaven. He described to me what was happening and what could happen and that I had been chosen for something important. Then, he said I had to understand the pain in the world before I could move forward. He would brush his hand across my brow, and I would literally feel the people's pain, anguish, suffering, and cries for help. It was heart-breaking and overwhelming. Every time I slept, I experienced more and more of the same. Every time, the emotional wave became worse."

Sam paused again. He remembered the pain, and it caused him to tear up.

"Please continue," Laura urged.

"This is where my story gets crazier. I ... I spoke with Jesus. He came to me in the dream. Jesus spoke to me about what might be to come. I would be a major factor in the future of humanity. Then he blessed me. I can only assume what we watched in the video is the response from that blessing."

"Anything else?"

"Jesus told me you were the one person I could trust."

Laura smiled at the comment.

"I must say. If I did not see what we watched a hundred times previous, I would have thought you went crazy. Anyone else would have you committed. However, there is something I never told you. I was told there would be a time when you would come to me for guidance. I was also told to give you your name.

"Jesus came to me when you were first brought to the orphanage. He told me you were sent to me for a specific purpose and that I needed to give you the tools to grow and mature in a way that would allow you the opportunity to choose the path he hoped you would choose. This is going to sound as crazy as what you just said. But you, Samson Elijah Jazeel, were sent by God to be his best hope for humanity."

Sam sank back in his chair. He slid his hands over his face and through his hair. "Well, at least I know I'm not going insane.

"What now?" Laura asked.

"Now I train my ass off. Jesus said he gave me talents, talents I will grow to master over time. I have to take some time to become proficient with them. Then, we'll determine where we go from there. I hope to make some progress by the time we talk a month from now."

"Great. That gives me time to follow a lead I have about a possible assassination attempt on the President."

"Us or a third party?

"Not us ... at least not yet. However, this might be the way we deal you back in the game. I'll brief you on more if the lead pans out. In the meantime, go train your ass off."

"I'm on it, boss. Until next time."

Sam clicked the terminal switch to shut it down. He leaned back in his chair and contemplated everything for a moment. He felt good that God chose Laura to be his guide or confidante.

He could tell she was a morale woman, and she had been chosen, like he had, for her part in the coming events.

It was nice to have someone who believed him and confirmed he wasn't insane. That was becoming a general worry for him until now.

CHAPTER NINETEEN

S am spent the following day cleaning the house from top to bottom. He wrecked the place during his time dealing with the painful dreams. Once he put the house back in shape, he decided to go for a hike in the woods.

There was a ridge that led off to the north of his house. He followed that ridge for about a mile until he came across an old logging road. The logging road was not easy to see with the decades of growth filling it in. But he was able to detect the outline of the road because of the older growth trees bordering it to both sides and the trail left from wildlife using the road for decades.

Sam decided it was best to stay off the trail. Predators liked to stalk the trails for prey. Sam didn't want to fend off a bear or, worse yet, a pack of Timber Wolves. Their numbers had exploded since the war. What he wanted to do was find a quiet place to meditate.

He came across an old-growth oak tree with a comfortable niche between two roots he nestled down in with his legs crossed. His head and back rested against the tree, and he listened to the rustle of leaves, the occasional falling branch or twig, or the occasional bird song or squirrel jumping from one tree to another.

A feeling of peace curled around him like a hug from a loved one. His mind flashed to the meeting with Jesus. The love and joy he felt during the conversation with Jesus could not be compared to any he experienced before. He always hoped the bible represented Jesus in a true light. The experience turned out thousands of times better than anything he could have hoped for.

Sam started meditating on what Jesus said about his blessed powers. His mediation was supposed to be the key to unlocking his new talents. He decided to think about time in reference to what Jesus said about slowing it down.

He closed his eyes. Some moments later, little flashes of light started appearing to him. The sounds of the forest started fading. When he opened his eyes, they focused on a leaf falling from the oak tree he sat under.

The withered red leaf twisted this way and that as the air caught it during its flight to the forest floor. As the leaf neared the ground, it seemed to hang for a second.

During that time, all sound ceased. The leaf paused a moment before it started back on its descent to the other leaves awaiting it on the forest floor. The sparkles he saw went away, and the sound of the forest flooded back to him.

"That was it?" Sam asked out loud.

Sam stood, walked over to the leaf, and picked it up. After returning to the house, he placed the leaf on the island in the kitchen he used to cut meat on.

While he cooked dinner, he kept thinking about the leaf and the way it hung in the air for a moment. The event happened so fast he started to think he imagined it.

He left the skillet where his sausage simmered and

grabbed the leaf. Sam closed his eyes and focused on the same thing he did when he sat at the base of the oak tree, the love from Jesus. When he opened his eyes and tossed the leaf, little flashes of light appeared like before and then, searing pain. He grabbed his head and collapsed to the floor.

Sam woke in a dream. The angelic Paul stood looking down on him in his kitchen. Everything was dark in contrast to Paul. It was as if a fog bank rolled right in through the front door and set up camp in his home. Sam stared up at him and blinked his watery eyes.

"Any chance Jesus sent you to deliver a user manual on practicing my new talents?

"I'm afraid not."

"What happened?"

"This is the result of not following the warnings that you were given."

"I couldn't even tell I had done anything!" Sam sighed and shook his head. "Guess this is the part where I'm supposed to rest?"

Paul smiled at him.

"You can injure yourself by trying to force the ability again too soon after doing it. The human mind is a frail thing and can, in essence, short circuit."

"Cute," Sam said. "How will I know when I can attempt one of the abilities again?"

"It would be safe to attempt once a day based on the current strength of your mind. With practice maybe more often." Said Paul.

"Once a day! That isn't much at all."

"Look at your powers this way. You are the one person in the world with such abilities. Your mind has been designed in a way that is a bit different than any other mind. You have memory abilities far exceeding any other person's.

Do you think it is a coincidence you acquired the ability to remember anything you see with perfect precision? God chose you for this long before your parents ever conceived you. Yet, it took the right choices to make the journey to this point. In this instance, the choice not to push yourself too hard when practicing your abilities will keep you sane. If you push too hard, you will drive yourself mad."

"Got it. No trying to be a superhero."

"If that is the way you wish to look at it."

"What now?"

"Now you will wake up with a splitting headache and need to deal with the grease fire in your kitchen."

"Greeeaaaat!" Sam rolled on his back and closed his eyes.

The pain hit him well before he opened his eyes. The smell of smoke and the crackle of fire brought him to his feet in a hurry. He rushed to grab a fire extinguisher to put out the grease fire threatening to spread from the stove to the wall.

Once he had the fire under control, he put on his sunglasses in an attempt to help with the headache Paul warned him about. They didn't work at all. Sam turned off the stove and dropped the fire extinguisher. Collapsing in a nearby chair offered no relief to the pain in his head. It pounded as if someone hit him in the head with a rubber mallet with each pulse of his heart.

"OK, time for bed," he groaned.

CHAPTER TWENTY

S am awoke the next day in a much better state. Though he was a little apprehensive about testing his time ability out so early in the day, he decided to practice his Wing Chun on his Muk Yan Jong training dummy he built out of oak.

Oak and pine were prevalent in this area of the Smokey Mountains. Pine wasn't near as strong or dense as oak. Sam preferred ash or teak for his Muk Yan Jong. However, oak worked fine for what he needed, and the hard wood gave him something solid to pound on during training.

With his dense bone structure and well-callused hands, he could destroy a wooden Muk Yan Jong training dummy. In recent years, he started using padded gloves. The padded gloves were an attempt to protect the training dummy, not his hands.

He didn't have any gloves at the house, so he trained with bare hands and worked technique and speed. Maybe on his next trip to town, he would attempt to locate some gloves. If he couldn't, he would purchase some cloth to make some hand wraps out of.

While he worked through his well-practiced moves, he started thinking about his parents being killed in cold blood. The anger kept rising in him as he worked away on his moves. Jake or his handlers had them killed.

He started imagining what he would do had he been there to confront those who were sent to execute them. At that moment, he started seeing the same sparkles and flashes in his vision as he did when he meditated in the forest where the leaf paused before it hit the forest floor.

Moments after that, cracking sounds emanated from the Muk Yan Jong where he slapped the offset arms. Sam paused, inspected the arms, and realized they broke at their mounting points. He had barely touched them, yet they splintered through to their cores. The arms were four inches in diameter at those spots. These were made of oak, a tough and resilient wood.

Sam realized he experienced the same sensation as he did when meditating in the forest with the leaf. He identified a trigger, the sparkles in his vision. He also figured out how to set off two of the four talents. The first by peace and love, and the second talent off his anger. The latter allowed him to snap two four-inch oak arms without a hint of extra effort.

The next day he felt more confident about resting periods and decided to try the strength test out on an old rusty iron plate an inch thick. He strapped the twelve-inch by six-inch plate to two stacks of cinder blocks. He meditated and kept at peace while striking at the plate with a hammer fist technique.

Sam hit the plate five times as hard as he dared and did nothing to the piece of iron except knock off some rust flakes. Then, he stared at the iron and thought of his parents getting tortured and murdered. The anger surged through him, and as soon as the sparks of light hit his vision, he struck with all his strength.

He heard a resounding "crack" like someone snapping a dry branch. The piece of iron flew off the stand and into the woods. A moment passed before he realized his forearm was the cause of the loud sound. The pain reached his brain around the same time. Sam grabbed at his arm and grunted in pain. Something snapped in his arm.

He'd never felt the sensation of breaking a bone before. He was glad this was the first time. He set the bone back in place as fast as he could with a scream. The nanites would start trying to repair the area, and if the bone wasn't set within minutes, they would create a bond at the site. The bone would be required to be broken a second time to set it in its proper place. The downside of his nanites.

After setting the bone, he went inside and scanned his arm with a medical scanner showing he had indeed broken the ulna bone, and he saw a hairline fracture in the radius. The ulna bone was the smaller of the two forearm bones, and it took the hardest of the blows to the iron plate. The bone was back in the proper place, and the scan showed it was starting to heal.

The nanites responded to the area and kept the inflammation to a minimum while providing the resources to the area to repair the bone. After a couple of hours, his forearm just felt sore. The majority of the pain was an afterthought.

Sam walked back out to find the piece of iron he struck. He located the iron plate in a pile of some old rotting leaves and pine needles it landed in. The iron deformed from the blow by about three inches.

"That is incredible," Sam said.

He turned the piece of iron over and over in his hands while smiling in wonder. Sam laid the piece on the table along with the leaf he tested his other power with a couple of days before. He sat down and stared at them for some minutes.

The two remaining undiscovered triggers still needed to be determined. He would need to go to town to figure out how the last two worked. Now that he figured out the sparkles he saw in his vision indicated the talents were active, he would be able to find the next two much easier. At least, that is what he hoped as he drove towards the largest town within a decent driving distance.

CHAPTER TWENTY-ONE

B efore the war, Utica, New York, maintained a healthy population of one hundred thousand people. After the war, it almost became a ghost town. Once the pandemic was cured, the town opened its schools back up, and people started migrating back to the small town.

Its population ballooned back up to around twenty thousand residents. Utica remained a fraction of its former size, but so did any city or town not deserted after the pandemics.

The town's board used the money from the schools to deconstruct the abandoned locations to bring down the crime rate and make the town a safer place for the schools and the students.

It took many years to demolish the old homes and relocate the few residents still in abandoned areas. Those areas were turned into parks or allocated space for growing schools.

Sam blinked in disbelief as he drove into town. Utica made Washington D.C. look like a cesspool, and a lot of money was spent on making that city appear clean and appealing. The big difference was the miles of abandoned buildings Washington D.C. retained. Most of the country resembled the outskirts of Washington D.C.

The leaders of this town must have their heads on straight. They did what needed to be done, and the town prospered because of their actions.

Utica was designed for everything to be within walking distance for college students. That also meant Sam would need to find a secure place for his truck for a few days. He drove down what appeared to be the primary road through town, and as luck would have it, a hotel rested next to a parking garage for short or long-term vehicle storage.

Sam pulled into the valet area and gave the truck over after locking the vehicle into valet mode, preventing the vehicle from being driven farther than a few miles before shutting down and alerting him.

He checked into the best room in the place overlooking Utica College. The hotel was twenty stories tall, making it the tallest building outside of the town's center. That gave him a fantastic view from the balcony. With his binoculars, he scouted out areas he wanted to use for testing his ability to blend into the shadows.

St. Luke's hospital lay to the west of the college campus. He might be able to test his healing ability there in some way he was yet to determine. He could not hide his excitement about attempting to heal someone.

Nobody died of disease or infection anymore. That didn't mean people stopped needing medical assistance. Sam's advanced nanites gave him the ability to heal within hours or minutes. The rest of the population still dealt with physical injuries and with diseases that were genetic in nature.

Sam took a shower, then dressed in some black designer jeans and a dark blue form-fitting long-sleeved shirt. He walked down the street to a local pub for dinner. It was a warm fall evening, and the college students were out enjoying the last frost-free days of the year before they bundled up for the winter.

Sam often forgot he was as young or even younger than most of these students. He acquired his doctorate before many of them thought about a bachelor's degree. He received a lot of looks from the young women who were out and about.

Sam was not a shy person—his training cured any hint of that. He always made eye contact and kept himself aware of everything else going on around him. His physical appearance might have changed, and his fingerprints might not match his original ones, but he was still a fugitive. The last thing he wanted to do was become sidetracked by a woman, even some of the ones as beautiful as he saw around town.

Sam found a pub down the street from his hotel. Being around people made him realize how much he missed Alyssa. He missed being around people in general.

Sam took a seat at the bar and ordered a big steak filet, a large salad, some seasoned wedge potatoes, and a house ale on tap. The atmosphere was great. Music blasted, and the place was packed with students.

The bartender was an attractive older woman around her late twenties or early thirties, Sam guessed. They struck up a conversation. He could tell she was attracted to him by her body language and the way she kept looking at him at every possible moment she wasn't scrambling to gather drink orders.

He thought about asking her back to his room. He'd

only had sex a few times in his life, and she had all the curves he liked. She was fit, witty, sexy, had a great butt, and legs that went on for days before they disappeared into her short shorts.

He chatted with her for a couple of hours after finishing his meal in between her getting drinks. As he was about to ask what she might be doing after her shift, a man came in and gave her a hug and a kiss.

When Sam realized she was attached, he spun around in his seat, faced the crowd, and nursed his third beer, enjoying the atmosphere. Sam couldn't help but revel in it. Smiles crossed his face more this night than in years of being undercover. Though technically he was still under-cover, he mused.

Thoughts of Alyssa crept back into his head. He wanted to ask Laura about her many times, but he was afraid the news he would receive would not be pleasant. So, he kept the questions locked away. It didn't stop him from thinking about her.

CHAPTER TWENTY-TWO

S am stayed until the crowd started thinning out near one in the morning. He decided the time had come to walk over to the hospital and find someone to heal. He hoped Jesus would guide him in this journey. He didn't have a clue where to start.

As he approached the front entrance to the hospital, Sam decided to be sneaky. He would sit in the ER and pray for those who might come in.

If Jesus guided him, he wouldn't need to worry about the timing. Jesus would send someone in needing help, and Sam would be there at the right time. Certainly, that was how it would work out.

He explained to the head nurse in charge of the ER he wasn't looking for anyone. He wanted to sit in the waiting area and pray for anyone who came in. She cocked her head while she thought about it and decided to let him stay.

The ER was outfitted as any small town's ER might be. This night, the doctors and nurses stood around with nothing to do. Sam decided to pray regardless. He said a quick prayer for anyone who might be hurt this night or any day coming.

He had mixed feelings as he left the ER a few hours later. With nobody coming into the ER, that left Sam still needing to figure out how to use his healing power.

He decided he would be blunt and come to the hospital tomorrow during visiting hours and see if anyone who might be terminal wanted to speak with him. He checked the sign near the door indicating the visiting hours and walked back to the hotel.

On his way back, he realized he was the only one outside. The empty streets were a bit unexpected, considering the bar was rocking a few hours before. Sam strolled along with his hands in his pockets, enjoying the cool night air when he noticed a foot patrol coming towards him.

"Shit," Sam whispered under his breath.

The patrol already headed in an intercept course. The realization dawned on him that curfew must be ongoing after his disappearance. He was, after all, a few hours from the capital. His former self was still wanted for attempted murder on Jake.

Four troopers made up the patrol. By their appearance, they were government patrolmen, not local police. They were armed with stun beaters. Every hit provided a forty-thousand-volt shock. That amount of voltage would knock most people on their butts, even Sam.

Sam's heart rate started to pick up. He didn't need to be questioned about what he was doing out this late at night. He didn't know a solid cover story. Telling the patrolmen he went to the hospital to pray would sound like a lame excuse to cover something else.

The patrolmen approached to within fifty yards. The street he walked had a row of trees giving great shadows from the streetlamps. Before they got within range to hail him, he slipped into the shadows.

Just as he got into the shadow, his vision started sparkling as it did with the other two powers he tested. His vision also lost all color. He started seeing everything in black and white. He took that as a sign he was hidden and his new talent active.

The patrolmen picked up their pace after they lost sight of him. Sam took his hands out of his pockets and moved in a practiced crouch. He floated along the tree line and up next to a building while the four men walked past the point where they should have made contact with him. When they realized he'd disappeared, they all pulled out their stun beaters.

"Where did that guy go?" one of the men said to the rest.

"There isn't an alley through here. He should be right here," one of the other patrolmen replied.

Sam kept moving past them and down the line of shadows. They pulled out flashlights and started pointing them around. By then, he moved well past them. He kept to the shadows until he reached sight of his hotel.

He waited around for a minute to make certain the coast was clear. When he confirmed no one was in sight, he stood, put his hands back in his pockets, and strolled up the steps to the hotel entrance. He nodded to the clerk behind the desk.

"Sir, don't you know there is a curfew in effect?"

"No, I wasn't aware," Sam replied.

"You're lucky you didn't run into any patrolmen."

"Oh, yeah? Are they a problem?"

The man's nose scrunched up in disgust. "They're assholes. They give you the third degree, and if you don't answer the way they like, they'll slap you with one of those damn stun beaters."

"That sounds unpleasant. Lucky for me, I didn't run into any while I walked back from the pub."

"You are at that. They love messing with young people."

"Thanks for the tip. I'll make certain I don't stay out this late for the remainder of my trip. Have a good night or morning, as it is."

"You do the same, sir."

Sam made it back to his room. He took off his shirt and walked out on the balcony. With his improved eyesight, he picked out the patrolmen. They'd given up looking for him and were walking down the road leading towards the campus and hospital. He would not have chosen that particular moment to try and test his shadow talent, but he was happy he'd sorted it out.

While showering the next morning, he realized healing someone in the ER may not have been one of his best ideas. That would have caused a stir. Maybe that was the reason no one came in.

Sam had other things to do other than testing his abilities. He needed to stock up on supplies. He decided to gather supplies prior to stopping by the hospital.

If what he planned to do succeeded, he would want to leave before word spread. Miracles of healing would be the hot item for the press, and the last thing he wanted to do was stand out. He felt like he had no choice.

The new plan would be to head to the hospital and find someone who needed his help. Then, come back in the dark of the night and heal them.

He spent the morning getting supplies and dry goods to stock up on. Once he packed all his gear away in his truck, he locked the vehicle back up and walked over to the hospital.

He entered the main entrance instead of the emergency room like he did the previous evening. A pleasant older lady sat at the information counter and greeted him as he walked up to her.

"Hello, sir. How may I help you today? Are you looking for someone staying here?"

"No, ma'am. I do not know anyone staying here. Every now and then, I like to come to a hospital and meet with anyone willing to share their story with me and pray for them if they will let me. I was hoping you might direct me to anyone who is staying here who would like to speak with me in that manner?"

"Well ..." she paused. "It's a rare sight to see someone of the younger generation caring about others in need. Most of those who grew up vaccinated think hospitals are for nothing more than treating physical or mental injuries. There are many diseases in the world nanites do not treat."

"There is always a price to pay for success. In this instance, our youth require a proper explanation that things were not always as they are now. It is up to those coming before them to uncover the facts about what happened in history and how not to repeat those mistakes."

The lady smiled. "Right you are, young man."

She shuffled some papers together on her desk and leaned forward.

"Listen, we do have some families here dealing with some unfortunate things. This hospital is tied to the college and has much better technology and care than many local hospitals. So, we receive the difficult cases."

Sam put his elbows on the table and leaned in towards her.

"Do you think any of those families would welcome the empathy of a stranger? I would be happy to meet with them beforehand so they can decide if they want to let me meet the rest of the family. I just want to help even if all I accomplish is listening to their story and sharing in their grief."

The lady got a sad look on her face.

"Son, I hope you understand what you are asking for. These situations are quite sad."

"I understand, ma'am. That is why I'm here."

"Let me make some calls. Take a seat over there, and I'll be with you if I find any families who would like to speak with you."

"Thank you, ma'am, I appreciate your time and your indulgence of my request."

Sam smiled at her and walked to the nearest chair and took a seat.

About fifteen minutes later, the lady came over to him. Sam stood to greet her.

"A gentleman will be down in a few minutes to meet you."

"Thank you."

Not long after, a man with dark, disheveled hair, sunken eyes with dark circles under them, and a slow gait came toward him from the elevator. He glanced at the lady behind the desk, and she motioned to Sam. The man met Sam's gaze, and he scrutinized Sam for a moment.

He walked over to Sam and shook his hand with a limp grasp as if he were too tired to bother.

"Can I help you?" he asked.

"Sir, my name is Sam Colt. I wanted to lend an ear or a shoulder if you would be so willing. I would love to hear your story or to sit with you and pray if you wouldn't mind."

"What makes you think I care what God thinks? He

doesn't seem to care my child is dying of an incurable disease."

"The fact you are pissed at God is an indication you wish he would save your child. Maybe you and I can go somewhere for a bit and talk in private."

The lady had been listening in and spoke up before the other gentleman responded.

"A private waiting room is located down the hall next to the men's restroom and vending machines."

Sam could tell the man was thinking the offer over.

"Shall we? I'll buy you a soda. If you do not want to talk, I can share my story first, and if you still do not want to talk, I will be happy to leave you in peace with your family."

Sam motioned towards the hallway. The man started moving that way as his curiosity won out over his reluctance to speak with this stranger. The two made the short walk to the vending machines, where Sam bought them both a soda and a small bag of potato chips.

Sam opened the door to the small private waiting room. The room had a wall monitor, a small table, four chairs, and a small love seat.

Sam took a seat across the table from the other man and let the man drink his soda and eat a few chips. The exhaustion on the man's face said it all. His gaze would often wander off as if he were looking at something else Sam could not see.

After a few minutes, Sam spoke to him.

"May I ask your name?"

"Edward, but everyone calls me Eddy."

"Interesting. My name is Sammy, but everyone calls me Sam."

"I still don't understand why you want to sit with us," Eddy said.

"Well, let's call it divine inspiration. I recently lost both of my foster parents. The loss made me ... To be honest, I sort of lost hope. Then, I realized others are dealing with the same pain as I or worse. I thought maybe some of them might want to share their pain. It is obvious you are exhausted from grief."

Eddy half-smiled.

"I suppose I do look pretty bad. Leaving the hospital is difficult with my little girl wasting away here. I need to be here if she calls out for me."

"What coaxed you out of the room to meet me?"

"I don't know. The lady up front called me and asked if I wanted to talk to someone or pray with someone, and I had at that very moment been cursing God for our misfortune."

"Why is it you blame God for your misfortune?"

"Isn't it obvious? He sits up in Heaven on his throne while we suffer down here like ants in an ant farm."

"Have you heard the phrase, Free Will?" Sam asked.

"Yea, what of it?"

"Do you believe you have free will?"

"Yes, I do."

"So, you believe in God, even though you are pissed at him. You also believe in free will. How can you be angry at God if you believe in free will? God gave us free will. That means we create our Heaven and Hell on Earth, not God."

Eddy sat for a moment, chewing on the words Sam just said to him.

"But why does he allow these things to happen to good people?"

"For the same reason a parent lets their children make their own mistakes—so they can learn from them. How many times have you seen a parent tell a child not to do something because some consequence will happen that might hurt them? How many times has the child still done the very thing the parent warned them not to do? God lets us live a life of free will, and that means we create the obstacles in our life, not Him."

"Are you saying humanity is killing my daughter?"

"Well, not exactly. I'm leading into this. God loves us all and wants to save us all. Because of the gift of free will, that is not possible. His plan is to save as many of us as possible from ourselves. To do that, he sometimes has to allow what we consider bad things to happen in order to save a soul."

"OK, I think I'm following you."

"Good. now let's say God cured your daughter. Would you live the rest of your life devoted to God? Would you have lived a life devoted to God had she never fallen ill? Could my presence here be part of his strategy to help me save you and your family?"

"But my daughter is dying in a bed. Hope is lost. The doctors say she is terminal."

Sam leaned forward. "Are doctors human?"

"Yes."

"Are humans fallible?"

"Well, yes, of course.

"Then there is always hope, and if the worst should come to pass, knowing she is in Heaven with Jesus at her side is a blessing in itself because she is no longer suffering."

Sam paused a moment.

"If you'll allow me, I would like to meet the rest of your family and pray with you for your daughter. If not, I'll bid you a pleasant day and leave you in peace to watch over your daughter."

"I think ... I think that might be nice. We closed ourselves out of everyone's life we used to hang out with. I think my wife and I would welcome someone to talk to besides each other. We ... we are so tired of the waiting and wondering."

Sam held out his hand, and at the same time, asked Jesus to heal this man of his exhaustion and give him back his energy to continue fighting for his child. Eddy took his hand and shook it again. A surge of energy flowed through him and into Eddy. Eddy perked up, and his eyes widened.

"Who did you say you were again?"

"Just call me Sam."

"Nice to meet you, Sam."

"Lead the way, Eddy. I'd be honored to meet your wife and child."

Eddy walked with Sam to the elevators and up to the fifth floor, where their room was located. Sam took note that Eddy had more energy since his prayer and shaking his hand. Eddy spoke of his wife and daughter while they made their way to the room.

They had been married two years before having their daughter, Ashley. When Ashley turned three, she started having seizures. Within a year, they progressed to the point where she needed to be monitored at all hours due to how dangerous they became.

She spent her fifth birthday in the hospital on constant care. The doctors gave her less than six months to live based on the progression. They believed the condition stemmed from an abnormality in her brain that managed her nervous system.

Sam and Eddy approached the room and opened the door to find his daughter was in the process of another seizure. Eddy ran to her, but the doctors prevented him from getting to her while they worked on getting her through the episode. Eddy's wife Trudy, a lithe 5'3" young woman, clutched for him and he grabbed and held her while she wept.

Sam bowed his head and asked the Lord to help her hold on a bit longer until he might position himself to help. Her seizure ended when one of the doctors injected her with something. Once they stabilized her, one of the doctors spoke with Ashley's parents.

"We increased the dose again. Every time we do, her heart rate drops lower. If her heart rate drops below the threshold of where her body can keep it beating, she will go into cardiac arrest."

Trudy started sobbing while Eddy held her close, tears streaming down his face. He nodded at the doctor to go away. As the doctor turned to leave, he eyed Sam.

"Who are you?"

"I'm a friend of Eddy's."

"Oh, I'm sorry I did not see anyone here with them the past few months."

"Not a problem, doc. You are doing what you can."

The doctor nodded and walked out of the room. The nurses tended to a few things to make certain Ashley was comfortable and soon left Sam with Ashley's parents.

Trudy, Eddy's wife, stopped sobbing long enough to realize that Sam was standing near Eddy.

"Eddy, who is he? Is he with the damn insurance company again?"

"No, no." Eddy calmed her with a kiss on her forehead. He grabbed her hand and pulled her over towards Sam.

"This is Sam. Sam, this is Trudy, my wife."

"It is a pleasure, ma'am. I'm honored."

Sam held out his hand as before, and Eddy, sensing by now there was something special about Sam, pushed Trudy's hand into Sam's. Trudy's eyes widened as Sam held her hand in both of his. She made eye contact with Eddy, and he nodded at her.

"Eddy explained to me the courageous fight your daughter has been putting up these past couple of years. It saddens me to see what she is going through firsthand."

Trudy just stared at Sam as he let go of her hand.

"I was wondering if you would allow me to pray for your daughter?"

Trudy and Eddy looked at each other.

"We would like that," Eddy said.

Sam reached out and grabbed both of their hands, then led them to the foot of Ashley's bed. Sam stood with Trudy and Eddy on either side. The frail body of their daughter lay on the bed in front of them. Sam brought to memory the love he felt when he sat with Jesus.

"Lord, I stand here with Trudy and Eddy as we look on as their daughter fights for the life you gave her. We are asking if you would bless her and give her the strength to continue the fight and win out over this malady threatening her life. In Jesus's name I pray, Amen."

After the prayer, Sam stayed and spoke with Trudy and Eddy until night had fallen. He learned how they met, how many hours Trudy spent in labor, what they did for a living, politics, religion, and many other things. What he needed was a few moments he could be alone with the child.

"Eddy, you two must be famished. I would like to offer to buy dinner for the three of us, if you would permit me?"

"We don't want any handouts."

"Eddy, this isn't a handout. This is one human being offering to provide food to another as a sign of friendship. Please?"

Eddy glanced at his wife.

"OK, then. As long as it isn't charity."

Sam got out some money and handed it to Eddy.

"If you wouldn't mind getting the food, I would like to talk with Trudy a bit more about Ashley."

"It's OK, hun. Go grab us some pizza from the cafeteria and some sodas."

"Pizza sounds delicious," said Sam.

"Alright. If you need anything, page me from the nurse's station."

"I will," she said.

Eddy shuffled off while Sam spoke with Trudy about where she grew up, how Ashley was when she was a baby, how he grew up with a photographic memory and remembered everything.

He spoke to her about how his parents were murdered before he'd turned one, and he wasn't fortunate enough even with his ability to remember much about them. Then, he excused himself to the attached bathroom. When he came out Trudy was standing at the foot of her daughter's bed.

"Phew, I don't know about you, but I did not realize how much I needed to use the restroom."

He sat back down, and they talked for a few minutes about this and that until he got the result he'd hoped for.

"Sam, I need to use the restroom and maybe wash up before Eddy gets back. If you notice anything wrong with Ashley, call for me or press the call button on the bed."

"Trudy, I'll watch over her like she was my own."

"For some reason, I believe you will. I don't know why ..." She trailed off as she got up and went to the restroom, closing the door behind her.

Sam waited for a few moments, slid out of his seat, and walked over to Ashley. He didn't have long before Eddy came back with the food from the cafeteria.

Ashley was coated in sweat. Her little body twitched with micro seizures even now. That might be a precursor to another major one. He felt so sorry for her in that state. Sam smiled knowing he was about to allow God to change their lives. He reached out and grabbed her little hand in his.

"Lord, please cure this child of her ailments and let her lead a full life free of disease."

Sam's vision started sparkling, and a surge of energy pulsed through him and into Ashley. Her pale color faded away into a warm pink. Her pulse jumped back up. Her breathing became steady. She opened her eyes to gaze upon him.

"Did Jesus send you?"

Sam chuckled. "Yes, he did, little one," he whispered back to her.

"Are you the one the angels said would come help me?"

Sam smiled. Jesus was ready and waiting.

"Yes, sweetie. You are going to be fine now. You'll be out of here in no time. Your parents will be here soon. Your dad is getting pizza, and you momma is behind that door cleaning up for dinner. They have watched over you for a long time."

"Thank you," she said, squeezing his hand.

"You are welcome, my dear. I need to go now. You have fun eating pizza and being with your parents."

Sam brushed the hair away from her eyes and leaned over and kissed her forehead. She lay back down and closed her eyes.

He stood back up and slipped into the corner of the room while generating the same feeling he did when the patrol tried to track him down.

The sparkles in his vision appeared, and everything faded to black and white. He was hidden in the dark corner of the room. The one remaining light was from the light above the bed. He hoped enough shadow remained in the room when Trudy opened the door that he would stay hidden. The toilet flushed about the time the door to the room opened, and Eddy came in with the pizza.

"Trudy? Sam?"

"I'm in the restroom, Eddy."

"Where is Sam? Wait, Ashley? Trudy! Ashley is awake!"

The door to the bathroom burst open. "She's awake?"

Neither of them seemed to notice Sam standing in the dark corner behind them. Eddy dropped the food he was carrying on the table and rushed to the bed with Trudy. Ashley laid in the bed with her eyes open, smiling.

"Mommy! Daddy!"

Both Eddy and Trudy burst into tears and hugged her. They had not spoken with her in weeks. She was supposed to be in a drug-induced coma. Eddy pressed the call button for the nurse's station.

Sam knew the room was going to fill with bodies and staying unnoticed would become impossible. While Eddy and Trudy were distracted, he slipped to the door and cracked it open just far enough to slip out. Neither Eddy nor Trudy noticed the movement of the door. Their entire focus was on hugging and kissing their daughter, who seemed to be in high spirits.

Sam made the short walk to the stairs in time to see the on-call nurse rush to the room. He smiled. God was amazing, and this gift was by far the most rewarding of the four blessings he received.

"Thank you, God," Sam said.

It was time to leave. The story would not take long to spread. The press would scour the town looking for him.

Learning how to use the power needed to be done. He sighed. Laura was going to be pissed that he did this. Again, it had to be done.

Sam made his way to the ground floor and outside, where it had gone dark once again. This time he wasted no time in going stealth. The last thing he needed was running into a curfew patrol for the second night in a row.

He figured out on the way back that when his vision was black and white, he was hidden in shadow. When he saw sparkles, his movement could be seen. When he started seeing color, he was visible to the naked eye. This knowledge allowed him to stay out of sight as much as possible on the way back to the hotel.

He checked out and went straight to his truck. He double-checked his cargo, powered up the truck, and headed out. Once he made it out past the last of the town's light towers, he turned off all his lights and let the vehicle drive itself until he was about ten minutes from his house.

While the autopilot drove, he checked the status of his house security. Surveillance video picked up three deer and some coyotes that tripped the motion sensors. Nothing looked out of place. He flipped on the lights and drove the rest of the way under manual control.

He pulled into his garage and carried in all his purchases made up of dry goods and other general items that should prevent him from having to make another trip for at least a month.

He sat on his couch and thought about the past week. It had been an eventful week. He figured out the signs for each of his talents.

The healing ability was a no-brainer and left little to learn. He became proficient in the shadow walk ability. That left the most difficult and powerful skills: slowing time and his unnatural strength. He would need to practice those once per day. Tomorrow he would start on trying to improve them.

Sam spent the next three remaining weeks improving his abilities while he waited for his next contact with Laura. He also started keeping tabs on current events in the news. His old face stopped showing up. Officials claimed he made his escape out of the country.

The news of the little girl who was healed by a stranger was all the rage on the local news stations. An artist's version of his identity was shown. Lucky for Sam, the rendering was not a good representation of his face. He felt better about that, though camera feeds did catch a decent image of him coming into the hospital. For some reason, the camera images were all out of focus. Lucky day or maybe someone higher up was at working helping him.

No man hunt had been established. He had not been involved in a crime. That left reporters trying to hunt him down, which would prove difficult for them due to his location.

By the time he was due to speak with Laura, he was able to improve his slow time talent to three seconds and his strength to fifteen seconds.

The progress he made on the strength feat pleased him. His shadow ability improved to the point where he could sneak up on predators and use it multiple times a day.

His favorite game was messing with the coyotes in the area. He would sneak up on them and smack one of them on the ass. They would jump three feet straight into the air. By the time they landed, he'd disappeared. Some would give chase for a bit, and that made it more entertaining for him. They always ran in the wrong direction. Then would stop and sniff the air.

He hadn't tested his healing ability since the first time in town. He didn't feel the need. His real frustration came from attempting to freeze time. Three seconds was a lot of improvement from less than a second. However, this gave him little time to achieve much.

He used the ability in conjunction with his shadow meld for scaring coyotes. Freezing time gave him about three or four steps of complete silence before time continued again. This allowed him to judge how much he could accomplish when using it. At this point, the ability would come in handy for an escape, in a fight to change the advantage, but not much else.

Still, Sam wasn't going to complain. It was, by far, the most difficult ability for him to use. He could kill three people in three seconds if they happened to be standing near him. That was an amazing advantage.

Sam didn't look forward to speaking with Laura about being on the news. He hoped she would get over it before they spoke.

CHAPTER TWENTY-THREE

S am walked the mile-long underground hallway to the safe house location. The past few months presented some of the most difficult days of his life. He thought about his parents, his mission, and his confidante, Laura.

She represented the closest thing to a parent he had left in the world. She was the one person in the world he put his complete trust in. Alyssa would make two. Unfortunately, Sam didn't know if she lived through the government purge.

After meeting Jesus, Sam's resolve strengthened tenfold. Revenge and hate no longer filled his thoughts. Judgement of his enemies would come through Christ.

By all accounts, the overthrow of the government would bring about serious change for the better that would not just change the United States of America but affect the entire world.

Countries would not survive without support from each other. That included the likes of America and other heavy hitters left in the world. As the tip of God's spear, he was determined to be as sharp as a spear tip could be.

Sam popped the door open on the far end and scampered up the ladder. He was running a few minutes late. Laura initiated the communication link before he popped up into the small room. She sat sipping some sort of hot beverage while waiting on him to arrive.

"Sorry I'm a bit late, Laura."

Laura turned to the screen as she took a sip of her drink.

"Not to worry. I'm just enjoying a nice hot chocolate. Living underground gets a bit cold."

"What have you been up to since we last spoke? You mentioned something about a plan and an investigation you were probing into."

"Yes, that is why I wanted to talk to you. But first, how is the training going?" she asked.

"Not bad. I'm progressing. I think I'm ready enough for anything you have in mind."

"You look much better. Your color is back, and you appear to have added some weight back as well."

"Yes, I can pack on muscle with ease. I can also lose it with ease, as you've seen. What is going on in your side?"

"Well, we've been watching an investigation unfold in a town not too far from you. Someone matching your description saved a young girl from dying. Authorities are interested ..." she paused and raised an eyebrow at him. "Care to comment?"

"Yeah, crazy stuff, right? The things reporters say to get airtime."

Laura kept staring at him. Sam glanced around for a few seconds with innocent puppy dog eyes.

"OK, it was me. I needed to test my talents!"

Laura rubbed her head.

"Didn't I tell you to be careful?"

"Yes, and I was. Nobody knows who I am."

"OK, you cannot predict the future or who you might run into. The last thing we need is for you to run into someone who recognizes you. We are going to have to speed things up."

"Why?"

"What happens when you require another town trip?"

"I'll drive to another town."

"Hmm, it's that easy then?"

"I get the warning. It won't happen again. In my defense, there is no other decent way to test that particular talent."

"Well, there's nothing we can do about it now. What's done is done."

Laura took another swig of her hot chocolate and picked up a piece of paper on her desk and scanned over it for a second before continuing.

"So, the other item on my list. We've been hearing rumors the President has been resisting the efforts of Congress over a treaty and a trade agreement with South Africa. "This treaty is chaffing him the wrong way. We need to understand more about the treaty and the trade agreement.

"We were able to acquire an audio recording between two specific individuals confirming the President is being monitored, and a plan is in motion to take him out if he vetoes the next attempt to pass the treaty."

"Who is the recording from?" Sam asked.

"Your old friend Jake Tillman was one of them. His role has changed to an enforcer for the inner circle now."

Sam's face turned sour at the mention of Jake's name.

"Because of the timetable, I'm sending you in to protect the President."

"What?" Sam exclaimed, his eyes wide.

"I know. He's on our list to remove from office. But for now, he's the enemy of my enemy."

Sam fell back into his chair.

"How am I supposed to protect him while being hunted by everyone in that location?"

"Your mission will be to infiltrate the White House and have an impromptu meeting with the President face to face. You will inform him of the situation and ask to be put on his personal protection detail or whatever he can to keep you close to him."

"Hmm ... I don't foresee any part of that being successful."

"I guess you'll need to rely on those slick new talents you've been trying to master." Laura winked at Sam.

"Well, I guess there is no point in trying to ease myself into the deep end. Might as well go for a cannonball and try not to get everyone wet in the process."

"Have faith, my son."

"Oh, I have no shortage of that. It's the failing the mission part I'm worried about."

"Now is the time to find out how great I always hoped you would become," she grinned.

"How will I communicate to you in D.C.?

"You won't, but we'll be watching. We have no shortage of eyes in that location. You are on your own for now. We'll keep our fingers crossed."

Sam said his goodbyes and shut down the safehouse, sealed it. He did the same to the main house before packing up the truck and driving toward D.C.

He planned to setup camp northwest of the White House within the capital's city limits. George Washington University was near the area and was about the only thing in use.

The suburb was called Foxhall Crescent. He located some abandoned homes near Glover Archbold Park. The park, along with the woods inhabiting the whole area, would allow him quick escape routes should the need arise.

The trip required an eight to nine-hour drive depending on road and bridge conditions. He gave everything in the house one last inspection, checked his gear in the truck, jumped in, and set off. He let the truck navigate the route while he assembled some electronic equipment he would need for his mission.

While his truck navigated the way over the rough roads and bridges washed away by floods years past, he pulled up satellite information on the White House grounds and the interior layout. He also checked the news to find information on the President's current location.

The President's itinerary listed him as out of the state at the moment. Sam checked the black logs he still maintained from his days watching over Jake. The President's schedule showed him back in town within a few days. That would give Sam time to find out how difficult infiltrating the White House grounds would be.

Sam arrived at the outskirts of the city about an hour after sunset. Once he'd made it to the neighborhood he was interested in, he shut down the lights and navigated by manual drive and night vision. He ran thermal scans in search of anyone inhabiting the area and found the neighborhood abandoned the same as he remembered when he lived in D.C.

Sam found a house in decent shape with a garage large enough to squeeze his truck into. The house was in shambles inside. Decades of looters stripped the home down to its skeleton. Not a single piece of furniture remained. He decided sleeping in the cab of his truck would be safer and more comfortable.

He slipped into his gear and set the timer on his watch. GPS estimated his time to the White House on foot to be around one and half hours. He expected about an hour if he moved fast and stayed to the shadows.

His approach to the White House grounds would be the General William Tecumseh Sherman Monument. Satellite imaging of the White House's south lawn presented the most cover along the east side of the lawn to the main building. Getting to the building without triggering an alarm would be the hardest to accomplish. Sam needed to investigate the south lawn to determine what obvious obstacles he would be up against in order to slip through the yard to the balcony. Two days remained before the President would be on site and another three before he left site again.

Sam made great time. He used the shadow ability, following the most overgrown areas, which allowed him to jog most of the way to the monument. He kept his gear hidden under his bomber jacket. The cooler weather prevented him from being soaked with sweat by the time he arrived near the south lawn.

Once he arrived at the tree line on the northwest side of the General Sherman monument, he went into shadow again and climbed the tallest tree near the statue and stayed put for a few minutes to see if any alarms were raised.

Nothing abnormal happened, so he slipped a headband on containing a high-resolution camera with a scanning logic looking for and highlighting magnetic fields. The camera would show all surveillance devices in the area. The one thing technology never overcame was the presence of a magnetic field. All electrical devices created one.

Initial investigation revealed that the south lawn was well lit near the fountain. However, a lot of shadow appeared near the White House.

Sam turned on the visor and zoomed out all the way. The outer perimeter fence threw off some field resonance. That meant the fence is likely used as the first line of detection. The fence might be set to detect vibrations or be touch-sensitive. That meant he could not climb over it. It might even be able to scan fingerprints. Sam needed to circumvent the fence without touching it in the process.

The next thing he detected was the fountain. As expected, the centuries-old fountain resonated a field from the electric pumps running the streams of water. But the field size signified something more than pumps, and no surveillance cameras could be located on the fountain. There could be anything hidden underground near it. He would need to keep a keen eye on that area when he made his move to the White House.

Sam stared up into the sky and adjusted the focus on his visor. This was the primary issue he wanted to identify—drones. Three drones hovered approximately eighty feet high. They must only be stationed there at night when they were not visible. He never recalled seeing them here during the day when he was stationed in D.C.

The drones gave him an idea. The military used drones on a regular basis to control local cameras that monitored the same areas the drones did. If the drones picked up anything abnormal, the cameras would focus on those areas.

This gave him his way through the outer security. He wasn't worried about getting inside once he made his way to the balcony. The security was designed to prevent penetration from the outer defenses. Physical security teams handled the inner defenses.

Once to the balcony, he would circumvent the electronic access. His knowledge of security systems and electronics would allow him to get past that layer. He didn't hold a PhD in the field for nothing.

Sam doubted he would be able to do anything further that night. He did not. possess the equipment necessary to hack the drones from this distance. He did, however, have the equipment back at his truck.

Sam stashed his visor back into a zippered pocket and went into shadow. He pulled another device out of his jacket similar to his visor, attached the camouflaged magnetic field scanner to the tree, and focused it on the area where Marine One landed to drop off the President when he returned to his truck. He would be able to link with the device and receive wireless updates as to the presence of his target.

He climbed down the tree and made his way back through the brush to his safehouse. The next day he assembled an air-powered rifle capable of firing a Wi-Fi-hacking chip he'd developed for Reboot.

The small chip could be attached via a magnetic or sticky round. He would need a sticky round due to the drones having very little ferrous metal in them.

Drones were no bigger than two loaves of bread strapped together, loaves of bread that happened to be bristling with sensors of all types.

To access the security control on the balcony, Sam fashioned a detector that would scan for the security control module for the balcony door. The scanner detected the strength of nearby magnetic fields and attempted to isolate any communication paths by wi-fi or induction depending on how the doors were being monitored.

The device he left on the tree to detect when Marine

One landed didn't alert him until the middle of the next day. He confirmed the information by finding a public camera pointed in the direction of the White House and confirmed the nation's flag flew atop the building, indicating the President resided within. Tonight would be the night.

CHAPTER TWENTY-FOUR

S ome hours after dark, Sam hid in the tree he was in two nights before. He recovered his detection device and slipped it back into a pocket. He took off his bomber jacket and tucked it in a shoulder bag secured to him.

He pulled out his small air rifle and his visor. He slipped the visor on and found his targets hovering in the same location as the previous night. He loaded a round and took aim. He let a shot rip and missed his target by a few inches too high.

"Shiiiit," Sam muttered under his breath.

He loaded up another round and adjusted his aim. He fired, and the round attached itself to the side of the drone nearest him of the three hovering above the south lawn. The drone detected the physical contact and went on alert for a few moments before it was overridden by the chip.

He put the rifle in the shoulder bag and turned on his wristwatch. After a few moments, he was able to hack into the drone and take over the cameras.

Sam confirmed the cameras and drones were not at the

correct angle to detect him in the tree with their infrared feeds. Though, he would need to disable them for the sprint up to the balcony. He set an override script that would loop their stream for thirty seconds while he made his sprint for the White House.

Sam scanned the vicinity for civilians one last time. The cool temperatures kept the pedestrians to a minimum. Sam didn't notice anyone in the vicinity, so he slid back down the tree, hid his shoulder bag near the brush at the bottom, and readied himself.

In order to make it to the safety of the shadows, Sam needed to clear the fence without alerting any security from the White House or triggering any alarms.

Sam pulled on a hooded mask to hide his face. He stayed to the shadows and triggered his override of the camera system and drones for thirty seconds.

Gritting his teeth, he pulled the anger from within to trigger his strength and ran at the fence. Just as Sam neared it, he leaped into the air with a grunt, clearing the fence by at least two feet and landing with a hard roll on the other side.

He came out of the roll at a dead sprint. The abnormal strength gave him a massive speed boost. His strength faded as he neared the garden on the south end where the shadows were thickest and found a dark spot to crouch in. His heart pumped hard, and his hands shook from the adrenaline coursing through his body.

He looked at his watch, and after a few seconds more, the cameras came back online. He would need to trigger the camera loop again when he left.

It occurred to him he would not be able to circumvent the fence the same way on the way out. There was no way he would be able to clear it, and he couldn't muster his strength again until he'd rested. He would think of something when the time came.

Sam waited for a few minutes until he confirmed no alarm had been raised. The rumors of vibration sensors in the ground must have been exaggerated.

Sam found a handhold on the underside of the balcony, using it like a mountain climber to pull his way around to the edge and up on the banister. He lowered himself down on one of the benches and inspected the area for anything out of the ordinary.

He expected to find the balcony rigged with sensor plates to feel if someone attempted to break in through the balcony. His instincts served him well. He determined each chair on the deck rested on the original decking. A raised portion existed everywhere else. The chairs and tables became the new floor he would use to approach the locked doors.

He slipped on his visor and located a magnetic field of what might be the control module for the access system on the door.

A wall sconce for lighting no longer in use hung near the location. He reached out and latched onto the light fixture with one hand. All he could do was pray the sconce held his weight as he braced against the wall with his feet and hung off the fixture, his feet dangling inches from the floor.

Sam held up his linking device to where he hoped the access control unit was, and after a few moments, a link light turned green, indicating he'd secured the device. A few moments later, his scripts cracked the door, and he was able to put his feet down on the balcony floor. Sam grabbed the door and slipped inside.

CHAPTER TWENTY-FIVE

The President finished up a meeting and went to the kitchen to grab a sandwich. The secret service agent followed him back to his bedroom and checked the room once and wished the President a pleasant evening.

The President sat on a couch in front of a wall, clicked a button on the coffee table, and the wall came to life with the local news. He sat for a while, eating his sandwich and watching the news until his eyes started getting heavy.

He shook his head to clear the sleep from his eyes and stood. He slipped off his shoes and started unbuttoning his shirt while glancing at a chair in a dark corner of the room. The chair should be over near the desk on the far wall. The President took a couple of steps towards the chair.

As he watched, the shape of a human started appearing before his eyes. The corner was dark. But there was no mistaking a person now sat in the chair. The hair on his neck stood on end. He shook his head and started to move toward a light when Sam spoke.

"Do not move, Mr. President."

The President leaped into the air out of surprise.

"Terry!" he yelped.

"Not wise, sir," Sam said

Sam clicked a button on his wrist, and a metal panel unfolded from above the door along the frame, and another set of panels slid from that section to meet in the center of the panel. This formed a solid barrier, preventing the door from being opened, and joined the door and frame to the structure of the White House. In effect, it turned it into a solid wall.

The President ran to the structure and tried to find a way to get through. He beat on the wall and yelled for the agent who pounded on the door from the other side.

"Sir, if you do not calm down, I'm going to calm you down."

The President turned and ran for the balcony doors. He made it halfway across the room when a sudden jolt knocked him off his feet. He lost the ability to stand for a moment and slumped to the floor.

Sam slid a stun gun into his pocket and zipped it up. He moved over to the President, lifted him up off the floor, and sat him in the corner chair he'd sat in moments before. The President opened his eyes when a hard slap hit him in the face.

"Sir, you left me no choice. Now I need you to listen to what I have to say. It may save your life."

The President looked him in the eyes at the comment and squinted at him.

"What do you mean by that?"

"Exactly what you think I mean. I have a few moments before they start cutting through the door or wall to get to you."

"How in the hell did you break into my room?"

"How is not important right now; why is important."

"What do you want?"

"I want you to listen to me for two minutes. What I plan on saying may or may not come as a surprise, depending on what you know about the congressional members you work with. We've uncovered a plot to eliminate you. I do not work with the secret service. In fact, they cannot be trusted. You need to understand what I am about to tell you and what you hear is not to be shared with anyone. If you do, you take your own chances."

"I'm listening."

Unintelligible shouts came from the other side of the door Sam sealed. A dull thud, thud, thud of something heavy hitting the door could be heard.

Sam glanced over at the door and back at the President. He pulled a small audio device out of his pocket and placed it on the bed next to the President and clicked a button on the device.

"This is why you need to heed my warning."

The President heard two men whispering about him. They spoke about the failed treaty attempts and the President blocking them at every turn. The time neared when he would need to be silenced. The President couldn't believe what he was hearing.

"Who are they?"

"Your fellow congressmen. Those you consider your peers or your partners in crime. Take your pick."

"Get serious ... Is this some sort of ploy to try and blackmail me or sway me to do something for you?"

"Oh, you want the truth? Fine, here is the truth. I don't want anything to do with you. I think you are a corrupt and dirty politician who believes in the phrase 'the status quo is the way to go.' I think you would drown a child if it would keep you in power. However, those who sent me need you to stay in power, for the moment, and are willing to risk everything in an attempt to inform you of the traitors who are working against you. My particular skill set might keep you alive if you are willing to work with me."

The President thought for a moment. His mind kept hearing the thuds coming from the other side of the wall.

"Say I believe you. Then what?"

"Once you set aside some time to read over the data embedded in this recorder, you will believe me. As you can tell by the noise coming from the other side of the wall, I do not have the luxury of waiting for you to read through all the data now."

"What do we do then?"

"Once you read the data and realize the danger you are in, you are going to need our help to keep you safe from your own people. You will click the link at the end of the package of documents, and that will open a secure channel to me. At that point, you will hire me as your personal assistant. Appointing me as a member of the secret service will get me killed and expedite your death soon after. If I am your new assistant, I can maintain a close distance to you to protect you from your own people."

A sizzling sound emanated from the wall, and a moment later, a laser breached the wall next to the door.

"I was afraid of that. Read over the material and contact me when you are finished. Remember, if I wanted to kill you, the deed would have long since been accomplished, and I would have disappeared. That should be sufficient evidence of my abilities."

Sam took the device and stuffed it into the President's

pocket. The laser cut through a majority of the wall by that point. Sam ran for the balcony as the secret service blasted through the wall with a mechanized suit. Agents secured the President and rushed him in the opposite direction.

By the time the agents made their way to the balcony, Sam was halfway to the fence and in shadow form, but he would be spotted as soon as he hit the lit area of the south lawn. He tapped his wrist, and the drone he'd hacked earlier rammed into the other two drones, sending all three crashing down onto the lawn.

The unknown issue with the fountain started showing itself. The fountain on the south lawn shifted, and three openings appeared around the base of the fountain.

Three mechanized dogs came out from under the fountain. Their satin-silver bodies gleamed in the light from the south lawn. Their whip-like tails lashed back and forth like an upset cat. They paused until their gazes came upon Sam's figure darting through the yard toward the perimeter fence.

Sam realized the hounds needed to be eliminated for him to make his escape. If they tracked by scent, they could track him all the way back to his hideout.

Sam reached the fence, and in mid-stride, he pulled his molecular sword from its sheath behind his head and slashed an opening in the fence large enough for him to slip through. That alerted agents on the balcony and on the ground floor. Sam heard the familiar sounds of heavy rounds being fired at him.

He focused hard until time came to a halt. He dashed through the hole in the fence and sprinted for the street still in shadow form. A few seconds later, the sound of explosive rounds hitting where he'd stood moments before reached his ears. He sprinted across the well-lit street in full view of everyone, with the mechanized dogs closing the distance.

The agents lost track of him once he made it to the shadows on the other side, but the hounds still closed the distance. That confirmed they possessed the technology to track him by scent, sight, or both.

He pushed his legs as hard as he could until the very last moment when he spun to cut the first hound in two as it leaped for him. The other two circled him, looking right at him. They must be equipped to see other ranges of the visible spectrum he could not hide from.

They didn't attack. They seemed to be trying to contain him until reinforcements arrived. Realizing this, Sam decided to force their hand. He feinted at the first hound and whipped a knife at the second. The blade skipped off the hard alloy of the hound and skittered away into the brush. But the sword slice that followed cut right through the head of the mechanized hound. That left the last one, which leaped at him in an attempt to latch on to his leg.

He jerked his leg out of the way just before the hydraulic jaws snapped shut with a clang.

The hound jumped at him again and missed. As it went by, he stabbed his sword through the metal spine of the hound. It tumbled into the bushes, damaged but not out of the fight. It came at him on two legs with its metal jaws snapping at him like a turtle. A final swipe of his sword dispatched the hound.

Precious seconds had elapsed during the quick scuffle with the dogs. He made certain to take a few more moments to chop the CPUs of the heads into a pile of shredded pieces to prevent the secret service from downloading anything to analyze, such as his scent for another pack of hounds to track.

He finished with that, grabbed the knife he'd thrown at the second hound, and continued sprinting along into the darkness with the sounds of sirens, police, and secret service agents arriving on the scene.

He was lucky to be alive. Full autonomous robots had been outlawed as part of the cease-fire from the last world war. If the government used robot hounds as protection, their use might point to something bigger on the horizon.

He needed to pass the information on to Laura and see if her team might be able to dig up more information on the metal beasts.

As a precaution to being tracked back to his hideout, he jogged around in random directions for hours and crossed the cold Potomac River three times before circling back around.

Sam relocated the truck and rigged some detectors to alert him if his location was compromised. Then, he waited for the President to contact him. The evidence he gave him concerning different levels of government and how large the plot was to eliminate him was too much for the President to ignore. After three days, his communicator went off, indicating the President found some time to contact him.

Sam clicked on his tablet while sitting in the back of the truck in the dark with his mask on, eliminating any chance of determining his location or identity. He ran some algorithms that traced the call and set up a firewall to detect any piggyback traces that might be trying to locate his position. After checking the security, he brought up the link.

The President's lack of sleep made him appear rather haggard to Sam.

"I read over your information packet a few times, and I will agree to your terms. But I would like to understand how this is supposed to play out. How do you plan on saving me from my own people? Some of those people, I consider my personal friends ... some lifelong."

Sam nodded. He understood what it felt like to be betrayed by a friend. "Sir, many of those people have been coerced by a much smaller group of officials. Those are the ones who are the danger. Your puppets on the hill are so afraid of losing their grip on power that they're willing to do anything to prevent that power from slipping through their grasp. I'd be willing to bet you've let people die so you can keep yours. Why should you be so surprised when it happens to you?"

The President's expression showed Sam his statement stung true.

"As far as how this is going to end, I cannot say for certain. The plan is to keep you alive long enough to find out the specific targets and bring them to light. What we do at that point will be determined by what they do or say."

"Very well. What now?"

"I'm going to send you a packet of information on a new assistant you want to hire. If for some reason the new assistant is detained or dispatched during processing, other agents will make certain you and the rest of your family does not live through another week."

"OK ... I'll get the information submitted to one of my people for processing. The process will take a few days for all the background checks and the like."

"Not a problem, sir. We've taken care of that. It will go without a hitch. In the meantime, try and rest up. The threat is still a ways away. We have time."

The President sighed, and Sam noticed him relax a bit.

"I will see you in a few days. Signing off," Sam said.

Sam clicked off the viewer and searched through the scanning logs for anything abnormal. Tonight, he would stash his truck at a new hideout and relocate to a permanent residence set aside for his new identity outside of D.C., close to Maryland. It was time to go back to being a nerd, or at least looking the part.

CHAPTER TWENTY-SIX

The next day, Sam received a call on his phone requesting he come down to the White House for photographs, fingerprints, DNA scanning, badge creation, and everything else needed to gain access to the grounds of the White House and the rest of the Department of Defense locations when traveling with the President.

Only authorized personnel were permitted to wear sunglasses or hats on site. His sensitivity to light meant he needed to wear something to protect his eyes.

Sam procured custom-tinted non-prescription contact lenses. This would allow him to function throughout the day without raising concerns.

This was a dangerous mission. There were some old habits and postures he needed to change, or else those who were watching him might determine who he was. Jake knew about his sensitivity to light, along with other habits he possessed. With more than his own life at stake, Sam could not take any chances.

He made it through all the in-processing items without incident. He chatted with the guy who processed him in and found out they both had the same hobbies. Well, at least Sam let him think they had the same hobbies. He made lying seem so natural, he could beat a polygraph test while being administered truth serum. His mind made him the perfect operative.

After a few days of processing, he was provided a tour of the White House grounds and brought to have a one-to-one meeting with the President in the Oval Office.

"Mr. President, this is your new assistant you requested, Phil Konsiki," the attendant announced.

"Thank you, Allie. You can go."

"Yes, Mr. President."

The President walked over and shook Sam's hand as the attendant left, closing the door to the Oval Office behind her.

"Mr. President. I appreciate the opportunity to work with you."

"It was my pleasure, Phil. Happy to have you aboard."

"Have you gotten any sleep?"

"A bit, yes. Thank you for asking."

"I recently dealt with some insomnia, so I can relate."

The President sat on one of the couches and motioned for Sam to take a seat on the couch across from him.

"I like to jump right in. Do you have any important meetings I should be aware of immediately?" Sam asked.

"I have a diplomatic meeting to attend next month with South Africa. We have a security briefing this weekend. I want you to attend."

The President leaned forward.

"Do you know when the attempt might be made?"

"Not at this time. Once I have a better understanding of your coming schedule, I'll have a better idea."

"Well, I suppose we should get you settled and show you to your desk. It's just outside the door."

The President walked with Sam to a desk already outfitted with all the items the President's assistant might require.

"Feel free to arrange the area however you like. Just don't infringe on the rest of the team's areas."

"Thank you, Mr. President."

The rest of the day, Sam went through the desk and shelves designated for him. He came across an audio bug that he disabled but left in place.

The next few days leading up to the briefing, he familiarized himself with the staff, the cabinet members, and the secret service team assigned to the President. He read background checks on all of them, crosschecking them with his own database. Nothing popped out at him as abnormal.

All the cabinet members hid dirty things in their closet. Most political leaders had things to hide. None of them had what Sam would consider intent. They were dedicated to the President.

Assassinating the President was no small feat. Sure, Sam had infiltrated the White House, but he was no ordinary human. To his knowledge, nobody else existed in the entire world with his capabilities.

The President rode in a limo that retained some of the same benefits as a tank. An advanced protection field, supplied by Reboot before the company fell, protected the limo from ranged attacks.

The armor of the vehicle was comprised of a composite, ceramic, and alloy sandwich designed to resist high-velocity rounds and absorb high-explosive rounds. The wheels and suspension mimicked the suspension on his truck. That fact was hidden by a hub cap bonded to the synthetic tire.

The vehicle projected high-intensity microwaves ten feet in front of it and five feet to either side to discharge improvised explosive devices buried in the ground. The limo even contained EMP rockets, EMP grenades, a smoke screen generator, and a missile defense system designed to defeat any type of missile fired at it. Many other life-saving options existed on the vehicle as well. Sam was quite impressed by the list.

Outside of the vehicle, the President was protected by a drone network that scanned for explosives and weapons. Explosives could be detected by their molecular structure, and weapons by their shape and high-power battery packs.

If a threat was detected, the drone could project a beam of microwaves at the perpetrator, which would make them feel as if their skin was burning. That allowed the secret service agents to close in to capture or kill the suspect.

This led to Reboot determining the President's death would have to come from within. Originally, Reboot's plan for eliminating the President was Sam. He was likely the only man in the world with the skill set to get close enough to the President to kill him. That changed after Jake turned on Reboot.

Sam sat reviewing all possible scenarios that he would use along with any that could be concocted by a would-be government assassin.

During his research, Sam determined his hack of the drones was investigated. The back door he'd used to hack in was eliminated, and all the drones had been updated with the new security fix.

A military-grade vehicle would have the best chance of taking out the President. To prepare for that threat, quick response teams sat ready to intercept all vehicle or air-based threats.

Sam knew the feds had something up their sleeve. Someone would need to take the blame for their actions, but the plot would need to thicken before Sam could discover their treachery. His first suspect was Jake.

After diving into his new role within the Inner Circle, Sam ruled Jake out as the actual killer. Though, he didn't rule him out as the orchestrator.

CHAPTER TWENTY-SEVEN

S
am arrived with the President in the Situation Room located in the West Wing of the White House.

The President greeted everyone and introduced Sam. The head of security and the lead secret service agent along with the Secretary of State, Secretary of Defense, and Secretary of Homeland Security attended.

"Everyone, this is my new assistant, Phil Konsiki. He now maintains my schedules, briefs me on meeting minutes, and maintains my general wellbeing. Please welcome him and give him a bit of slack while he learns the ropes."

Everyone greeted Sam and shook his hand. They all took their seats while the head of security gave him a briefing on the coming meeting.

"Mr. President, as requested, the meeting is being held in the National Archives Building. The meeting location will be a reminder to South Africa that our history was rife with civil rights issues. Our Bill of Rights will be on display in the very hall the meeting will take place."

He flipped on a viewer on the wall, showing the route the limo would take to the National Archives building.

"We will exit the White House grounds and proceed to 15th Street and south to Pennsylvania Ave. That will take us on a direct route to the Archives building. Alternate routes Bravo will use Constitution Ave and Charlie will use G Street over to 9th Street and south to the Archives building.

"We will enter the building from the main entrance, where you will meet with the South African diplomat and be led on a tour of the facility. The meeting will take place after the tour. When the meeting ends, you will exit the building from the side entrance, and one of the previous three routes will be taken to rendezvous back to the White House grounds."

Sam examined the routes. The President glanced at him, and he gave a little nod of assurance that everything appeared fine.

"Sounds simple enough."

"Brian, any change in the state of affairs in South Africa?" the President asked the Secretary of State.

"No, sir. South Africa managed to retain the whole of the interior of Africa after the collapse of the Central African governments post pandemic. The country's new borders range north as far as the Sahara basin. NATO voiced concerns over the land grab, yet NATO has no real say. No government bodies exist to protest the land grab in the areas taken by South Africa.

"This, in essence, makes South Africa a world power based on land mass. A treaty or trade alliance may bring stability to the region and the world. However, the remaining NATO nations have voiced concerns over South Africa's civil segregation and unrest. The segregation issues have been going on for some years now."

Sam interrupted.

"Actually, South Africa's segregation and discrimination issues go back to the very beginning, when the area started as a ship provisioning location for Dutch ships during the spice trade. The Dutch brought in slaves to help maintain the site.

"The British annexed Cape Colony in 1806 after the Dutch East India Company went into bankruptcy to keep it out of French control and to refit their ships on long voyages to their other colonies in Australia and Indonesia. In the mid to late nineteenth century, diamonds and gold were discovered inland. In the years since, wars erupted between the local tribes over resources and land. No civil laws were put into place when the British took over. This led to over one hundred years of a wealthy white minority in control of a black majority.

"In the 1990s, segregation laws were abolished due in part from international pressure and from internal protests that, on occasion, became violent. This led to a change, but it was slow. World War III set back any progress made in the country due to a power shift to a political party led by a white supremacy group who wanted the old era segregation brought back.

"With every other country of the world focused on recovering from their own ordeals, this allowed the South African government to go about business as it saw fit to do. That makes the treaty talk between the U.S. and South Africa governments a delicate matter due to the nature of the South Africa government continuing to segregate its majority native African peoples from the minority whites. The world has recovered enough that governments are starting to watch each other once again."

The President glanced at the Secretary of State, who

was a bit red in the face after the lecture from what he considered a peon assistant.

"Still learning the ropes?" asked the Secretary of State.

"Easy, Brian, he didn't mean anything by that."

The President couldn't hide a smirk on his face that belied the fact he enjoyed seeing the Secretary of State get instructed like a schoolboy.

"I'm stating facts here. I'm not trying to insult anyone," Sam said.

The Secretary of State cleared his throat.

"The treaty between our two countries is going to hinge on what the South African government is willing to do for us. We might be able to overlook their segregation issues if they are willing to give us something in return."

The President leaned towards the Secretary of State.

"Explain to me, one more time, why do we give any care in the world as to what South Africa is doing in their part of the world versus us in our part of the world? This isn't pre-World War III, where the cities are packed with people, land is at a premium, food is shorthanded for many countries, and there is a constant chess game going on with the major powers to make certain they hold on to that power?"

"Mr. President, as I just stated, South Africa has aggressively expanded to acquire all the land south of the Sahara Desert."

"Yeah, so? We acquired all of Canada when they fell. That has been common practice by most of the remaining countries."

"South Africa also holds the best robotic scientists and tech in the world. Our military is running on fifty-year-old tech with very little new technology introduced in recent years. This puts us at a disadvantage in the coming decades it would take to catch up to South Africa.

"In the meantime, they are building a new military more advanced than ours. If we don't get a treaty in place soon, our ability to make war will be severely diminished. They are poised to become the new world power. Had we been able to get the nanite tech from Reboot, we would have had a serious counter for their tech and the rest of the world.

"Thus far, we haven't had any luck with that. Though, I've been told Congress is working on a method to get us that tech. We can't move forward, assuming that is going to be successful."

"That explains the treaty a bit better. I'm still not going to bend to their will, Brian. As of now, we have the bigger military might. They might want to remember that when approaching me in the coming meeting."

Sam shook his head and kept his mouth shut. He wasn't here to be political. He was disgusted that politics always ended up being about power and money.

Unethical things were swept aside, or a blind eye turned so long as someone received compensation for ignoring their internal moral voice.

His mission was to protect the President and dig up information on the specific congressional members plotting the assassination attempt. That is what he decided to focus on. He did his best to ignore anything else said that didn't affect his ability to protect the President.

Nothing of further importance came from the meeting. The routes for the trip to and from the diplomatic meeting looked straightforward. If a plot to assassinate him was going to take place in the White House, Sam's earlier infiltration had most likely ruined any attempt that would happen there.

The Secret Service took enough heat from that alone.

Another breach in security would bring other agencies in on the investigation, along with a lot more unwanted attention to those behind the scenes.

The meeting adjourned, and Sam went on an errand for the President to drop off something to a congressman. Sam rigged up the meeting in an effort to meet with Congressman Gorden Linden of Illinois.

Reboot brought on Gorden as a legitimate inside contact, and Sam needed a purposeful reason for showing up there. The government capital remained a fluid environment where everyone became a potential threat. Everyone was under scrutiny, and Sam could not leave anything to chance.

CHAPTER TWENTY-EIGHT

S am went to Gorden's office to drop off a security terminal linked with his own through a Reboot satellite. It was the same link he used to communicate with Laura. Reboot took precautions after Jake turned on the company.

Sam showed up at the congressman's office and introduced himself as Phil Konsiki, assistant to the President. Gorden took the package and opened it. On the face 'lightning in the sky' appeared that the congressman was instructed the informant would give him. He thanked Phil for coming down on short notice and shook his hand.

Later that night, Sam made a secure connection from home in a cleanroom to Gorden, where Gorden informed him via text the attempt would happen soon.

The Circle of 21 coerced, bribed, or threatened anyone they needed to get their ideas and desires pushed through.

Even the President was included in the list. However, Gorden informed Sam that the group had grown much bolder as of late.

A few of the 21 started talking about breaking the

international treaty that prevented the research, development, and production of autonomous robots and machines. Drones remained the one caveat the treaty allowed.

With all this information, Sam started forming a picture. The hounds Sam was attacked by during the White House encounter broke from the treaty. The lack of news on the breach made sense now. He pulled up a copy of the treaty and studied it.

After reading the treaty documents, he realized the future arms race would be autonomous robots. This time it would be a race to seek the smartest AI for autonomous robots which would fight the next war. Without the threat of a plague to thwart them this time, the U.S. government poised themselves as the dominate world power.

Sam gritted his teeth. Politicians never learned or didn't give a damn in the first place. Their own country was full of empty cities and overgrown towns, yet they wanted to develop new weapons of war and take over new areas of the world.

That made the diplomatic meeting important. If the U.S. formed a treaty with South Africa, that would open up the U.S. to expand and take over the rest of North America, also expanding into South America while South Africa took the whole of Africa. That would make those two governments the world's new superpowers. *How long until they turned on each other?* Sam wondered.

CHAPTER TWENTY-NINE

S am spoke with the President the next day and asked him about his stance on the South African treaty. The President, a minority himself, felt that unless the South African government made major civil rights adjustments, he would be unable to accept any treaty proposition from them.

"Mr. President, this is the reason you are being selected. You are holding up a larger plan at expansion."

"What? That's preposterous. We already have tremendous amounts of unused land we need to recover. Why would we want to expand at all?"

"That's the nature of power and greed. If you think about it, the theory is sound. The U.S. is one of the best positioned countries. If the U.S. were to expand now, there would be nothing and no one to stand in their way. Two things are holding the process up. The international treaty against autonomous robots and an artificial intelligence able to function and control a mobile weapons platform.

"War is coming. It may not be here tomorrow, but this violation is a reaction to one thing: lack of bodies. Nobody has a stronger naval fleet than the U.S. What remains? The bodies to fill the ranks. That problem can be overcome if the treaty on autonomous robots is ratified or eliminated."

Sam sat back. He made his point.

"This makes no sense. I've backed them on everything they've ever wanted. Now they want me dead on one issue?"

"Yeah, that part we understand. Why do you think I'm here? This is an important step for them. And as a politician, it isn't what you have done for me. It is what are you going to do for me now?"

The President sighed.

"I'm getting too old for this shit."

"Don't roll over just yet."

"I'm not rolling over. I'm pissed off."

"Being pissed off isn't going to make killing you any harder."

He gave Sam a flat look. "So, now what?"

"Well ... I suspect the outcome of the meeting at the end of the month is going to be the turning point on if they pull the trigger. No pun intended."

"I guess we will see what happens."

CHAPTER THIRTY

T he days came and went with nothing out of the ordinary happening. Sam spent most of the time investigating the grounds of the White House.

He found his security clearance did not allow him full access to the premises as the assistant to the President. However, he was able to penetrate the closed-circuit network of the Secret Service and poke around in there for a bit, looking at anything they might have on the President or Reboot. He didn't find anything of interest to his mission.

The day of the meeting at the National Archives, the President seemed nervous. He called Sam into his residence.

He was dressed in a navy blue suit with a bright blue tie his wife had picked out for him. A white silk shirt gave a change of texture to the ensemble. Black oxfords rounded out the outfit.

"You look very presidential, sir," Sam said as he walked into the dressing room.

"I feel as nervous as a kid in church."

"I'm not going to say you shouldn't be. If you find your-self between me and someone trying to kill you, do not freeze, stay moving. Here, put this on."

"What is this, a belt?"

"It is a PPF."

The President examined it for a moment. "I've never seen one so small."

"The PPF has a charge time limited to ten hours, but it will work a lot better than the bulkier PPFs."

"I don't like wearing them. They present a weak image." He frowned.

"That might work in our favor if your own people think you do not have one on."

The President slipped off his normal belt and replaced it with the PPF belt Sam gave him. Sam pointed to a touchpad on the back of the belt clip.

"Press this to activate the PPF as soon as we're outside of the White House."

"I will."

The Secret Service arrived a few minutes later and escorted the President and Sam to the presidential limou-sine. They headed off down the planned route they were briefed on all the way to the front entrance of the National Archives building.

They entered the building, and the President greeted the South African diplomat. The diplomat appeared to tense when the President held out his hand. *This is going to be interesting,* thought Sam.

They were given a tour of the facility and sat down to eat after the tour. They chatted about the status of each country, world news, and the weather. Then, the meeting switched gears and delved into why the diplomat was in

Washington D.C. The South African government wanted a treaty and trade agreement. However, they did not want to change their ideals to get it.

The President refused to agree to any treaty terms unless they settled their civil rights issues and quit segregating other races in an attempt to keep the power with the minority whites.

The two got up and shook hands. The South African diplomat held the President's hand tight and drew him close to whisper in his ear.

"I'm sorry we did not agree on this matter, Mr. President. Maybe after the next election, our countries will be able to come to terms."

"Not if I have anything to say about it."

"Don't worry ... you won't," the diplomat replied as he released the President's hand.

The two turned and walked away from each other. The President received an escort back to the side entrance where his limo waited. The diplomat left through the front entrance.

Sam slid into the limo, and the secret service agent closed the door behind him.

"I suppose we should be thankful a brawl didn't start," Sam said.

"That bastard is a racist. If he is any indication of what the rest of the country is like, then I don't want anything to do with them."

The intercom rang in the limo, and the President accepted the call. The Secretary of State came on the screen.

"Sorry I wasn't able to make the meeting, Mr. President. How did it go?"

"Brian, don't get me started. Your ass should have been

here. I will not agree to any terms with those people while I'm the President. He talked down to me the entire time. For what, because my skin is darker than his? I'm the President of the United States of America!"

The limo started down the road. While the President yelled at the Secretary of State, Sam realized the limo did not turn on Pennsylvania Avenue like it was supposed to. The third alternative route was north of here. The Secret Service might be taking that route.

Instinct kicked in, and he turned on his PPF. Sam slipped his hand to the small of his back where the hilt of his sword rested. He mounted the scabbard upside down to hide the weapon in his suit.

The limo went up 9th Street to G Street, then turned west towards the White House. Something wasn't right. Sam looked behind their vehicle, and his hair stood up on his neck as he realized no escort vehicles followed behind them. No police escorts were in front of them either.

When they turned north on 13th Street, the street was dark. No streetlamps lit the roadway in this part of town. The President was still yapping at the Secretary of State about the meeting when Sam reached over and closed the connection.

"What did you do that for?"

"Shut up for a moment and listen. It's happening."

The President's eyes widened. "Oh shit ..."

"We have no escorts, and the front driver is not responding to comms."

The President hit the button to lower the front privacy screen to the driver, but the window did not respond.

"Shit." The President started looking around, frantic.

"Calm down. If we lose our cool, we're as good as dead."

Sam pulled out a small device, set it on the floor, and turned it on.

"OK, that should scramble any transmissions within fifty feet of the vehicle."

He took off his jacket, exposing the sling around his chest holding the scabbard in place on his back.

"Are you carrying a sword?"

Sam huffed. "Yes. Can I answer any more obvious questions for you?"

"What are we going to do? I don't want to die."

"Me either. I have no desire to die while serving a government official."

The limo came up on I Street next to Franklin Square. Not a single person walked the park with the temperature so close to freezing.

The limo slowed and turned into an alley between the Franklin school building and an office building with an underground parking garage.

In the alley, Sam could see four agents, two to either side of the narrow lane. They carried hypervelocity submachine guns. They were not expecting the President or his aid to be wearing PPF devices. The vehicle slowed to a stop between the four men who just stood there, weapons at the ready.

Sam assembled a small device and laid it in the floor of the limousine to block communications. He pulled his sword from his scabbard and closed his eyes for a moment to focus. The President was freaking out. Sam hushed him again.

"They cannot get to us with the doors closed. This vehicle is designed to protect you. So, calm down before I knock you out."

Just as he finished saying that, the vehicle shut off. An

electric motor sound emanated from the front cabin, and the front privacy window came down about two inches. A nozzle slipped through the crack, and the window closed back up to hold it in place.

"Shit, hold your breath!"

Sam jammed his sword through the section of the door retaining the hinges and pushed down with all his might. The sword cut through the door with some effort.

He pulled the sword out and shoved it through the latch of the door. The two agents on that side of the vehicle pushed against the door to prevent it from opening. Sam let his rage envelope him as he kicked the door as hard as he dared.

The door ripped off the car and took the two agents with it, crushing them against the wall of the office building, killing one and injuring the other.

Sam grabbed the President and pushed him out of the car. The remaining agents rushed around the back of the limo. Sam calmed himself, and time slowed, then stopped. The President was looking at Sam, and in one instant, he disappeared.

Sam leaped over the back of the vehicle, cutting off one of the agent's heads and cutting the second agent in half before time resumed.

The driver of the limousine moved around the front of the vehicle. He peeked around the side and saw the President cowering behind the front tire. He stood up and took aim with a rail pistol and fired off a shot. The tip of Sam's sword erupted from his chest twisted and slipped back out. The shot ricocheted off the President's PPF and struck the injured agent disabled by the door in the head, killing him. The agent who Sam impaled fell to the ground face-first.

Sam determined after a quick inspection that the limousine was not safe to drive.

"I suppose it's a good thing you're still alive. Wait ... ugh. What is that smell? Did you soil your pants?"

"Yes! I was scared shitless!"

"Wow. You *are* spineless. Let's move. It's a few blocks to the White House, shitty pants or not."

Sam pulled the President to his feet.

"We'll be walking right up to the front gate. Let's hope we aren't shot dead on the spot."

"Why did my own men try and kill me?"

"Chances are they were under some sort of blackmail."

"How do you know all this?"

"To be blunt, I'm a genius, and I am a trained operative. I know every street, every alley, and every underground entrance and exit in this city. Now, less of moving your lips and more of moving your feet."

Sam grabbed the device he used to block communications, broke it apart, and tossed it down a sewer drain. He removed his sword scabbard and harness, ready to throw them away as well, but paused. He had very rarely ever been separated from his sword. It was like throwing away a friend.

He had no choice—he knew they would be thoroughly searched for injuries when they arrived back to the White House.

He watched it slide off his hand and disappear into the darkness of the drain.

He pulled on his suite jacket and rolled around on the ground a few times to dirty himself up.

"Let's go before we freeze."

The two of them made the trip to the front gate without further incident a little later. The guards couldn't believe their eyes. They let the two of them right in. The President went straight to his residency and hugged his family.

Sam was escorted to a briefing room where he waited patiently for what he knew was coming, a very confused group of agents attempting to figure out how the President made it back alive.

Soon after, two secret service agents followed by Jake Tillman enter the room. Jake had a scene report in his hand he was looking over.

"Evening, Phil," Jake said.

"Jake Tillman? You aren't Secret Service. Isn't this a Secret Service investigation?"

"Typically, yes. Considering the circumstances and the nature of the scene we found at the limousine, I've been brought in as a special investigator."

"Where is the FBI? Isn't this their jurisdiction?"

"We are in communication with them. For the time being, they are allowing me to lead the investigation."

"I see."

Jake motioned to one of the Secret Service agents, and they walked to the wall, turning on a display for Sam to look at.

"What do you make of these photos of the crime scene?"

Sam spun around and saw the gruesome scene of the dead agents lying around the President's limousine.

"I don't remember much from the scene."

"How's that? You and the President were the only ones to leave it alive."

"The President and I were traveling back from the meeting with the South African delegation. The limousine took the northern return route, and while the President was speaking with the Secretary of State, I reviewed some agenda items we have for tomorrow.

"Next thing I know, the limo turns down an alley. We tried hailing the driver. The front window came down a few inches, and last thing I remember is seeing some sort of vent being wedged into place. Everything went dark after that."

"You didn't see anyone attack the agents?"

"As I said, no."

Jake popped up a picture of the door cut away from the limousine.

"Can you explain how the door was cut open from the inside of the vehicle?"

"No."

"What was the first thing you remembered after coming to?"

Sam leaned back and closed his eyes in thought.

"I came to outside of the limousine near the President. The agents were dead. I got the President to his feet, and we ran out of the alley, got our bearings, and made our way back to the White House."

"You don't have any idea how the agents were killed?"

"Something sharp? I think I remember one of them having been shot in the head."

"We know that. What I'm asking is, do you have any idea who killed them?

"No."

"You see where we are here?" Jake asked as he slid along the conference table, closer to Sam.

"Agents are dead. I'm sure their families will want answers," Sam responded.

"I want answers too, Phil. We all want answers. You see, someone came into that alley and saved you and the President. Based on the information we've received from you and the President, we have zero suspects."

"I would want to know why the limousine went into that alley in the first place."

"The President said the same thing, and we are investigating that angle. The problem we have is that all the witnesses blacked out, and the potential suspects are all dead. No evidence of anyone other than those people accounted for already entered or left that alley. That leaves one potential option."

"Which is?"

"One of the witnesses is a suspect."

Sam nodded. "Ahh, I see where this is going. You're looking for a scape goat."

"We are looking for the truth."

"So, you're implying that I somehow cut the door of the limousine off and killed all the agents without so much as a scratch to myself. Also, that I rescued the President and don't to admit it? Good luck getting a judge to believe that."

Sam stood from his chair, and Jake got off the desk to look down at Sam.

"I can't place it yet, but you seem familiar to me, Phil. You're certain we've never met someplace before?"

Sam just looked back at him. "Are you going to hold me for anything?"

"I could. But I won't."

Sam walked to the door.

"I hope you find what you are looking for."

"You can bet on it, Phil."

Sam started to open the door.

"Oh, one last thing, Mr. Konsiki. Don't stray too far from the city. I need you to remain nearby in case I have further questions about the investigation."

"Sure thing, Mr. Tillman."

Sam left the room.

Jake gathered the report off the table and turned to one of the agents.

"Put a full surveillance team on him. Eyes on, 24/7."

"Yes, sir."

CHAPTER THIRTY-ONE

T he next morning, Sam sat in his apartment, watching a live stream of a press conference the President called for.

During the press conference, the President stated an attempt on his life had been made by his own Secret Service agents. He gave his resignation to protect his family.

Sam sat upright from the couch.

"Son of a bitch."

He rubbed his face in frustration and gave out a growl. He stomped from the living room area to the kitchen and poured himself a glass of water. Sam leaned back against the counter and crossed his arms, thinking of what he should do next.

Sam went upstairs and pulled a long, slender, titanium-hardened case from under his bed. He presented eye and fingerprint identification, along with a voice command.

"Unlock".

The case clicked. Sam opened it slowly to reveal another molecular sword, the sister to his other sword he had to leave in the sewer drain.

He was always trained to keep backups to important gear. His sword was like an extension of himself.

Sam gingerly slid his hand around the hilt and pulled the sword from the case. He pressed the power button on the hilt and heard the power supply kick on, engaging the molecular edge of the sword.

He closed the case and pushed it back under the bed. He pulled an extra scabbard from the closet and slid it into place with a click. His mind went back to the scenario the President just placed on him.

Leaving would flag him as a suspect and bring the whole group of alphabet boys down on top of him ... again.

He could hear the news coming from the living room. The President would fly to Camp Humphreys and stay there for the remainder of the month while his resignation was finalized.

Sam walked back down the stairs as an idea popped into his head. He stopped at the coffee table where he put down his glass of water on the way to his bedroom.

He picked up the glass, chugged down the rest of his water, and went back upstairs.

Sam got dressed into a nice suit, put on a coat to protect from the cold, and grabbed a ride to the office of congressman Gorden Linden.

He made his way to Gorden's office in the Rayburn House office building and met with the secretary. She sat behind an antique mahogany desk at one of the bigger offices set aside for long-term congressional members.

"Hi, Stephanie, is Gorden in?"

"He is. Give me a moment, and I'll announce you, Mr. Konsiki.

"Thank you."

Sam took a seat and waited while Stephanie lightly knocked at Gorden's office door and went inside.

The office building, along with many other governmental buildings, were damaged back during the world war. The entire capital was renovated in order to present a strong look to the remaining nations that the United States was resiliant, even though that was not entirely true.

The country remained stronger than most. That wasn't necessarily saying much in a time when the population hovered around the tenth the size it was prior to the war.

Sam was admiring the architecture of the building when Stephanie popped back out of Gorden's office.

"You can go in now, Mr. Konsiki."

"Thank you, Stephanie."

Sam rose from his seat and walked through the doorway to Gorden's office. A twenty-foot cathedral ceiling with skylights letting in the natural daylight hovered well out of reach. The office didn't offer a large amount of space. Gorden made the most of it.

Shelves lined the walls crammed with books, documents, and bills to be reviewed or ones actively being worked on sat stacked wherever there was space.

Gorden sat at his desk that faced the window looking out over the capital. The interactions Sam had with Gorden in the past had all gone well. Gorden had provided some great information to Reboot during the lead-up to the betrayal of Jake.

Luckily, Jake was not kept in the loop on what contacts Sam made during his tenure as his head of security Gorden's anonymity remained intact.

The plump fifty something congressman popped out of his seat to greet Sam. His short beard and combover hair, greying heavily from years of tireless public service, or so Sam liked to think.

"Hi, Phil! What brings you to my neck of the woods?" Gorden thrust his fat hand into Sam's and gave it a squeeze.

"Gorden, I assume you've been watching the news?"

Sam motioned to the hovering eighty-inch video screen in the corner of the room.

"Aren't I always."

"I'm resigning my post immediately with the President, and I would like to see if you have room for me here?"

Gorden's expression changed to curiosity.

"So, you aren't going to aid the President during his remaining days?"

Sam shrugged. "He has no need of my talents, and I'm not allowed to leave the capital while the investigation continues. I hoped to get on with someone familiar, who already knows my skillset."

"I can definitely use the help. Though, I'm afraid I can't pay you what you currently make."

Sam gave out a laugh.

"I wouldn't have come here if money were the issue, congressman. Just trying to help serve my country in any way I can."

"Welcome aboard then, young man. I'll have Stephanie write up a transfer packet for you and get some of your information. We should be able to get everything settled today and have you start in the morning, unless you want a few days to settle your nerves."

"That sounds fine. I'm just eager to jump back in and help where I can."

Gorden ushered Sam back to the office door, and soon enough, Sam was back out the door headed home. He was thankful that Gorden had read between the lines and brought him in. It would help to stay busy and keep his cover intact until he could slip out of town and back to the safehouse where the truck was stashed to get in contact with Reboot.

He knew with the surveillance he was under right now, it would be too dangerous to attempt until the heat was off.

CHAPTER THIRTY-TWO

fter Sam was hired on with Gorden Linden, Reboot continued pulling in other employees hiding from the government and putting them in their dark facilities to protect them and bring their skill base back to the company.

They didn't stop with Reboot employees. Regular citizens with skills Reboot needed were recruited and brought under the protection of Reboot as well. However, the primary focus was getting military recruits. They recruited the lowest enlisted personnel all the way up the chain to Generals.

They had a lot of success in this area. Once Reboot recruited a few Generals, they started playing around with the divisions those Generals commanded.

Reboot assured them that the military would be used for containment and not asked to take part in the removal of government officials. The citizens of the United States of America were going to make the final call as to what effect the government would continue to run under, not the military.

Meanwhile, Sam and Gorden continued to research the twenty other members who ran the government behind the scenes. Out of that group, they narrowed down those members to around six who made most of the major decisions. The Group of 21 went by their lead.

Sam and Gorden spent a lot of time researching the rest of the congressional members in an attempt to understand who maintained a moral compass and which congressional members maintained their positions by coercion.

For those they found to be moral, they attempted to place others to protect them and report anything out of the ordinary. This is when Sam found out Alyssa Giovnia was still alive.

Other than Sam, she was the one other active agent from the class who still lived. She assisted and protected another congressional member who happened to be a retired General of the Army, General Joseph Nathan Forest. He still held sway over people in the military and was an important piece to recruiting military members for Reboot. Alyssa spent her time protecting him and transmitting information back and forth.

Sam took comfort in knowing she lived. He ached to see her, though he dared not. Any attempt to do so might endanger her position due to the suspicion he was under during his time with the President. Plus, he didn't look like the Sam she knew. She might try and kill him, thinking he was some sort of government spy.

Those who wanted the President dead still wanted questions answered about how the President survived the attack.

Sam had made certain none of the attackers survived the incident. Their knowledge was limited to the fact that the President's aid and the President somehow killed five well-trained men with a very sharp weapon, likely a molecular sword.

That left Sam as a large question mark. Sam knew they would make contact in the future or attempt to kill him–probably both. That left him no choice but to continue the charade of being Phil Konsiki, avid gym-goer and new aid to Gorden Linden.

A larger chess game was being played on a world scale, and he did not have a seat at the table yet.

Time would tell.

CHAPTER THIRTY-THREE

W eeks had gone by since Phil Konsiki and the President of the United States walked out of the darkness on a cold fall night to present themselves to the gate guards at the White House.

Jake Tillman sat looking at surveillance footage leading up to the attack on the ex-President for what felt like the hundredth time. In an effort to prevent condemning evidence, no drones were monitoring the actual attack.

Jake had argued if something went wrong, they would need the footage for review. In his mind, they could always delete the footage later. Out of fear of discovery, the majority outvoted him. His task now was to find evidence about what happened in the alley. In typical government fashion, the finger-pointing started as soon as the mission failed.

In the long run, they still got what they wanted. To preserve his life and the life of his family, the President had resigned.

Yet here Jake sat, looking over useless video streams. The limousine pulls into the alley, and half a dozen minutes later, the President and his assistant jog out towards the White House.

Jake knew the President couldn't fight his way out of a wet paper bag. That left the assistant as the suspect. He killed the agents and thwarted the attempt on the President's life. How he accomplished the deed was the question.

By reviewing older video data and studying the way he walked, acted, and held himself, he must be ex-military of some sort. If he were some nerdy assistant, he was the first nerd Jake ever heard of who spent more time in the gym than he did reading.

Jake sat with a Secret Service analyst looking over the footage and discussing the situation.

"He must be protection. The President must have brought him in after the White House was infiltrated. Was it a coincidence he got protection after an alleged perpetrator spent enough time in close contact with him to kill him ten times over? That meant they must have spoken about something ... but what?" Jake said to a video analyst.

"All great questions, sir."

"Yeah, and they were rhetorical."

Jake's conclusion was that someone figured out the attempt on the President's life. It was the only explanation that made any sense.

"So, this guy, this Phil Konsiki, is the key to the mystery. He stayed with the President and his family in the bunker until the President decided to resign. Then he gets hired by Gorden Linden, one of the 21. Gorden's background is one of inaction. He never makes a point to bring attention to himself. He never asks questions, argues against any positions, and never goes against the majority."

Jake paced back and forth in the small room. The analyst just watched him and kept his mouth shut.

That brought Jake to one conclusion. The assistant might be a mole. He may be the one who leaked the information about the attempt on the President. Jake would need some serious evidence to prove Gorden was a mole. There was the possibility he was incorrect. His new assistant, Phil Konsiki, raised all kinds of red flags.

Maybe Reboot planted him with the President. He decided an impromptu meeting would give him more of an idea on how to proceed.

The next day, Jake suited up in his normal attire, an expensive suit made of soft blue suede. Highlighting the ensemble were the two molecular short swords he carried. He took two security agents with him to Gorden's office.

Jake was well known around the area as an enforcer now. He gave up his position in politics when he gave up Reboot. He didn't care for politics anyway. It was too much bullshit for him to wade through. He preferred to be the muscle. Jake had chosen to stay with his undercover identity. Nobody but certain individuals at Reboot knew his true identity, and they were all in hiding or dead.

When Jake entered an area, people noticed. Nobody outside of the 21 would look him in the eye.

Once he'd made a few examples of those who thought they were outside his reach, coercing the other congressional members to make them agree to terms outlined by the 21 became easy. Members liked to talk tough on the phone or in emails, but when Jake showed up in person, things changed.

He walked into Gorden's office like he owned the place. Its rich woods and classic design signified this was an office of the United States government.

The room, which had been bustling with activity a few seconds before, went quiet enough to hear a pin drop. He looked over at the secretary.

"Gorden?" he asked.

She didn't say a word. She just pointed towards his office.

Jake took a moment to give the room a once over. He thumbed through a stack of papers on the secretary's desk.

"Maybe you should all take a ten-minute break."

The six employees filed out without a word. Jake stationed the two guards at the entrance with orders to not let anyone in until he'd finished his meeting with Gorden.

He walked to his office and opened the door. Gorden sat at his desk with Sam, going over some documents with him. They both spun around in their chairs when they heard the door open, thinking it was his secretary, Stephanie.

"Jake, what brings you here?"

Jake and Sam made eye contact. Gorden sensed the instant tension between them.

"Should I leave, Mr. Linden?

Jake interrupted before Gorden could respond.

"No, you will stay," Jake demanded.

"Why does this Mr. Tillman go and do whatever he likes?" Sam asked Gorden.

"He works for us as an investigator."

"Well said," said Jake, pointing to his nose and then to Gorden.

"I have a few questions for you both. I'll keep it short.

Please sit, Jake motioned to a nearby chair. Sam remained standing next to the Senator.

Jake took off his swords and sat them next to a chair as he took a seat. He opened a small tablet and hit a button on it to record the conversation.

"In an effort not to butcher your last name, may I call you by your first name?"

"Certainly," Sam replied.

For a moment, Sam worried Jake figured out his identity. He realized if he had, he wouldn't be here asking questions. He became very aware of the absence of his own sword no longer resting in the small of his back as he glanced at the two molecular swords sitting in their scabbards at Jake's feet. Sam kept his cool and played his role.

"You were with the President on the night of the assassination attempt, correct?"

"You know I was."

"Can you describe the events once the limo pulled into the alley?"

Sam sighed.

"As I stated that night in the briefing room, I don't recall anything other than the first few seconds. The front cab window rolled down a couple of inches, and someone wedged in some sort of vent. After that, I cannot remember anything. The doctors said it was my brain's response to the stressful situation."

"What is the first thing you remember after the vent in the window?"

"Uh, I remember being outside of the limousine next to the President. Agents were lying next to us dead. My first thought was to get the President to safety. So, we ran."

"You saw no one else in the alley?"

"No, sir. Just the dead agents."

"Why didn't you grab any of their weapons before fleeing? You stated a moment ago you wanted to get the President to safety."

"I didn't think of weapons. I'm not a trained soldier. I grabbed the President and ran."

"You appear like you work out. You are what, one-eighty, one-ninety?"

"Yes. What does that have to do with weapons training?"

"Phil, I'm an investigator. I'm trying to obtain as big a picture as I can based on the perception of others. Not an easy task, let me tell you. If I seem a little forward, I apologize. I've been trained to do what I do. I read not just the answers given to me but also body language, eye contact, facial expressions, and many other things to determine the information I need to lead to the correct answers."

"OK," Sam responded without blinking an eye.

"See, I know the President. I studied his background. He's your basic sissy. His whole life has been focused on being a politician. Politicians do not waste time going to the gym or doing weapons training. They hire others who do that for them, like me. The President did not kill five trained agents who were intent on killing him."

"I think I see where you're going."

"Maybe, but let me finish. The other person who ran out of the alley was you. We tracked no other operatives in or out of the location at any point on video. No other ways existed to leave the alley without being spotted. That leads me to the conclusion you saved the President.

"Further evidence suggests you do, in fact, have some sort of training. The way you walk, the shape you're in, your eye contact, and the way you refuse to sit down when speaking with me all suggest there is more to you than you are letting be known.

"Now, you might just be modest and don't want the attention, but maybe … just maybe you might have another agenda. That's the part I'm interested in. I want to know. I need to know what that agenda is because I'm paid to. Do you care to comment?"

"I have a few things I can comment on."

Sam paused as he feigned thinking.

"I grew up in an athletic family. When they were killed in the pandemic, I focused more on being in shape and healthy. My foster parents also instilled that mantra on me. To imply I have to be something more than an athletic nerd is a stereotype. In your mind, are athletic nerds are not allowed to exist? If there are athletic nerds, then some sort of training must be involved, leading them to be something more than they are.

"I do not remember any more than what I spoke with you previously about. If you have a problem with my answer, I don't care. Even if you did prove I was lying, there is no law against saving a human life."

"Interesting response."

Jake turned to Gorden. "Mr. Linden, how did you come upon hiring Phil here?"

"I spoke with the President not long after his resignation speech, and I mentioned I needed another person for my group due to our workload. He said nobody was better than Phil when it came to crunching data. Since the President was resigning, Phil would be out of work, and the President suggested I hire him on before someone else snatched him up."

Not bad for on-the-spot, thought Sam.

"Good enough."

Jake pressed the button on his tablet and typed in a few things before tucking it back into his inner jacket pocket. He stood and strapped his swords back on.

"I'll be keeping tabs on you, Mr. Konsiki. If I butchered your last name, I don't care at this point."

"If I may ask a question of you?" Sam asked.

"Please."

"Why do you wear two swords over a suit? It makes you look sort of ridiculous."

Gorden snorted back a laugh he couldn't retain, playing it off like it was a cough.

"Because I want to be seen. Everyone here understands who I am and what I do."

"If that's the case, then why the need for the swords?"

"You'd be wise to shut up now, Phil."

"Very well, just curious."

"That may be. Do you know what curiosity did to the cat?"

Jake turned and walked out, slamming the door behind him.

Gorden gave Sam a concerned expression after Jake left the room. The heat was on, and Jake was the fire. If Jake didn't find out his true identity, Sam proved he was enough of a threat that he would likely be watched even closer now.

However, he couldn't leave Gorden after just being hired. That would only put Gorden under more suspicion. Sam decided he would have to let the theatrics play out. He would focus on being a great assistant to Gorden until Reboot was ready for the next stage.

CHAPTER THIRTY-FOUR

J ake made the trip back to his office and listened to the
audio recording he took from Gorden's office.

Nothing about Phil Konsiki made sense. He
remained the one calm person in the entire office. Even
the front office became nervous in Jake's presence. He
remained calm and appeared confident and unflinching at
Jake's questions.

Jake knew without a doubt Phil saved the President in
the alley. Yet, he did not possess the evidence to prove it. He
pulled up the data from the scene where the President's
limo was recovered. All the secret service agents but one
was killed with a razor-sharp melee weapon.

The exceptions were crushed to death by the car door of
the limousine or took a bullet to the head. Jake did not find
any explosive residue in the vehicle to suggest an explosive
removed the door that crushed the agent. The door hinges
and latch showed signs of being cut by something very
sharp on either side.

The weapon used on the door must be a molecular

sword or something similar. That would coincide with the rest of the wounds on the agents. One agent was found in two pieces, and another lost their head by a single strike of the same weapon.

Molecular swords were not easy to manufacture or come by. Sam led the team that invented the latest iteration of the weapon, but this guy looked nothing like Sam.

Though, he did have the same physical profile, and his height was similar to Sam's. Could this be Sam in disguise?

Simple plastic surgery would not work. The facial recognition systems at all government facilities would have picked up his bone structure regardless of what his face looked like. If this was the same man, then his changes had to be achieved through nanite technology. He thought for a moment.

"What if ...?" he trailed off.

Jake took the audio recording and pulled it into a voice recognition utility. The utility analyzed the voice track from his conversation earlier in the day. He pulled up old footage of Sam when he was protecting Jake Tillman off an old database. An audio track of Sam and Jake talking before one of his speeches allowed a speech analyzer algorithm to compare the two tracks of audio.

Jake realized while listening to the audio clip from Sam that he did miss their talks. Sam had been the closest thing he'd had to a brother. But things were different now. They chose different sides.

If this person turned out to be him, he would have to kill him. He endangered the entire plan. Plus, if Sam was active, then Reboot still pushed forward with their plan of overthrowing the government. They might have developed a backup plan.

The analyzer came back with a result of 87.3 percent chance it was the same voice. To Jake, it was enough to sentence a man to death. He picked up the phone.

CHAPTER THIRTY-FIVE

S am returned to his apartment at the end of the day and popped open a fifth of tequila. His metabolism processed alcohol so fast that getting a buzz from alcohol required enough to put most people in the hospital.

It had been a stressful day. It took all his willpower to restrain himself from killing Jake every time he saw him. He caused the death of his foster parents and hundreds if not thousands of Reboot employees.

He stayed up and flipped through TV stations while finishing the bottle of booze and starting a second. He thought about Laura, his foster parents, Jake, his biological and foster parents, and Alyssa. When he finished the second bottle, he went to bed.

His wristwatch woke him during the night. He flipped on the display, and the slim holographic timepiece showed an intruder alert on the outer perimeter of the apartment. He still wore his button-down shirt but no trousers or shoes. A noise came from the other room, and he went into shadow and slipped out of bed. An agent of some kind entered his room wearing night vision. Sam reached him just as the agent spotted him.

Sam led with a strong palm jab to the throat. That kept the agent from calling out. The agent tried to bring around his weapon, but Sam disarmed him by twisting the agent's wrist followed by a forceful punch to the elbow, breaking it in two. Sam gave another jab to the throat, followed by six fast punches to the jaw. He kicked the suffocating man into the path of the next agent coming through the door.

That agent got a strong kick to the groin, causing him to scream out. When Sam realized his cover was blown, he grabbed the man's chin and back of the head and snapped the head backwards with all his strength. A loud crack came from the agent's neck as he fell to the floor.

Sam heard movement coming from downstairs. He snatched up his PPF and sword and put them both on. The first agent at the top of the stairs lost both of his arms.

Sam initiated his time stop long enough to dispatch the two others on the stairwell before they had any clue where he was. He made it to the living room at the bottom of the stairs to find two other agents standing near the front door where they came through.

He went into his rage and kicked the first agent so hard it stopped his heart and knocked him and the other agent through the front window and onto the front yard.

Jake sat in an unmarked van a block away, watching all this unfold through the camera streams the agents carried via contacts in their eyes. He had difficulty making out what was going on. What he did know was that his agents were dying.

"Smoke the house," he said to the tech sitting next to him.

A drone hovering nearby fired a small concussion

missile. Sam ran for the kitchen in an attempt to escape the strike. The missile hit and blew him through the wall and into the back yard. Sam wobbled to his feet. He was bloodied but not broken.

As he gathered his wits, rounds started coming in from the yard behind his. His PPF deflected the rounds while he sprinted for the next yard and leaped over the fence and into three agents on the other side.

They didn't see him right away because he was still in shadow. A couple of leg sweeps and boxing the ears of one of the agents disoriented them long enough to allow him to keep running on. Another explosion almost knocked him off his feet again. This round came from a large-caliber sniper rifle from a nearby building.

The tech turned to Jake.

"Sir, he is outside of our quarantine zone."

"Then chase him down and kill him."

"We are having a hard time tracking him."

Jake shot the tech a look. "Don't give me excuses. Just do it."

Sam slashed his way through the next two fences in his way as he kept working down the row of condos. At the end of the row was a drainage ditch that ran into a tunnel. That led to the maintenance tunnels that followed parallel to the subway tunnels that ran throughout D.C. Once he made it to the maze of maintenance tunnels, he was able to lose his pursuers.

He was again running for his life. This time he carried his sword, PPF, dress shirt, and underwear. He was battered, bruised, and bleeding. Luckily, the injuries were flesh wounds, and the nanites in his body would take care of in mere minutes.

He made for his secondary hideout where his truck

waited. He stashed fresh supplies and established an escape route back to his location in upstate New York. He could hide out in his truck for a short time while contacting Laura and figuring out the next steps. He did not anticipate Jake coming after him so fast. Something must have given him away. He would ponder that later. For now, he was freezing in the cold November air. Nanites didn't counteract cold.

The run to his truck took about twenty minutes. He huddled down behind some trash and rubbish and listened for any sirens or drones. All was quiet. They lost track of him for the moment. Once again, his shadow ability proved invaluable. He thanked the Lord for his gifts and moved into the building where his truck sat.

Jake walked through what was left of Sam's apartment. He went through the video footage from the attack and could not explain what he saw. He knew Sam was skilled in martial arts and with his sword. But there were moments during the video playback on the stairs where he disappeared and reappeared in another location. Plus, his strength seemed augmented. He knocked two agents through the front window with a single kick.

Two agents survived the attack. One headed to the hospital to get his arms grafted back on, and the other was the second of the two kicked through the front window.

Jake rifled through Sam's belongings. His apartment was bare. For an electronics genius, he carried a couple of tablets and a wall monitor that was destroyed in the fight. It looked like he kept his secret things at another location. Sam set this apartment up as a front. There was no doubting he was the same man he knew before, yet he remained different. Jake turned to the tech he spoke with earlier.

"Let me know if you find anything."

"We'll find him, sir."

"Oh yeah, keep telling yourself that. He expected us."

"We'll send out drones to watch for him."

"I need to report this to my superiors. If you discover any good information to report, do so immediately, no matter the time of day. It might save my ass."

"Yes, sir."

CHAPTER THIRTY-SIX

Dawn approached by the time Sam made it to his truck. He had less than an hour before the sun breached the horizon.

His window passed for getting out of town until nightfall. He cleaned up and checked his wounds. Some bruises remained from getting blown through the kitchen wall. They would all be healed within the hour.

He dug out some clothes and food from his truck and ate while keeping tabs on the news and monitoring the perimeter.

Drones flew past his hideout during the day while he waited for the cover of darkness. They moved too fast to pick up anything on their scanners. They might be hoping he would spook from the sound of them and come out of hiding.

CHAPTER THIRTY-SEVEN

J ake walked into Texas Senator Bryan Lambert's office. Senator Lambert was talking to an aid and looked up when Jake walked in. He motioned for Jake to follow him back to his private chambers.

They walked in, and the senator pulled out his chair and nestled into it. He motioned for Jake to sit as he grabbed a cup of coffee from his desk that still had steam rolling off it. He took a sip and let out a sigh.

"So, what's this I heard about you blowing up a house last night?"

"As you know, I've been investigating the penetration of the White House as well as who saved the life of the President. I concluded both incidents were orchestrated by the same person. That person is the President's personal assistant."

Lambert nodded. "I see. How did you come to that conclusion?"

"It wasn't easy. It took a lot of hours of investigation. The key was when I made a connection between the assistant, Phil Konsiki, and Samuel Creedy, my old counterpart from Reboot."

"How did you make the connection?"

"With this ..."

Jake pulled out his small datapad and laid it on the senator's desk. The senator picked up the device and played the two audio tracks.

"What am I listening to?"

"Those two different people are Phil Konsiki and Sam Creedy. When I ran a voice match against the two, it came up eighty seven percent positive."

"So, you are saying you tried killing the prior assistant to the President of the United States with a thirteen percent chance he was just some nerdy assistant?"

"No. With all the other data I pulled and the questions I asked when I met with Mr. Konsiki, I believe that Mr. Konsiki and Sam are the same person. If you add the results of last night's raid, you get a very skilled operative able to elude us yet again with a new set of skills I've yet to figure out."

The senator looked down at the datapad again and raised an eyebrow. "Can you extrapolate anything out of this mess?"

"Nothing concrete."

"Take a guess."

"Sam would not have come back without cause. I think that cause is to finish what Reboot started. We know Reboot went underground, but they're still operating out of many other countries. I think they're still trying to bring down the government. We may be at the starting stages of their plan and not know it."

The senator took another swig of his coffee and laid his head back on his chair for a moment and thought.

"Alright, let me speak with the others. We'll work on getting the national guard mobilized, and if need be, we'll make it legal to mobilize the Marines and Army to prevent anything from erupting across the country. In the meantime, you need to mobilize the police, Secret Service, CIA, FBI, and your security detachment to hold down the D.C. area. We can try and convince the military to mobilize and set perimeters outside of the D.C. area."

"Do you have any idea to the whereabouts of Mr. Konsiki, or is it Sam, might be now?"

"I believe him to still be within the city limits, waiting for sunset before trying to escape."

"I'll give you access to a military drone if that will help. Contact the CIA and tell them to establish satellite coverage in the area. If you can locate him, they can tag him and track him no matter where he goes. This has gone on long enough. End it."

"Yes, sir, I'm on it." Jake got up and left the office.

CHAPTER THIRTY-EIGHT

S am decided his best chance to contact Laura would be sooner rather than later. He managed to route a secure signal out to their satellite and contact her.

Laura sat at her desk when his call came in. She flipped on the secure monitor and could tell right away something was amiss by Sam's worried expression and remnants of wall insulation stuck in his hair.

"Laura, my cover is blown. We need to initiate the next stage of the operation. I saved the President, but that created the same problem as though he were assassinated."

She nodded. "We listened to his resignation speech. It was pretty pitiful. I thought he was going to cry."

Sam scoffed. "You should have seen him in the limo when the attack went down, blubbering like a scolded child. What can you tell me from your end?"

"When we realized he was removing himself from office instead of what we'd hoped he would do, we initiated stage two. Rallies started in all the major states concerning civil rights, jobs, city development, and the martial law brought down in D.C. with no explanation as to why.

"When we leak data on the corrupted officials, along with cells of people we positioned at those rallies, it should escalate the demonstrations and turn them into riots. We have informed our contacts in the military that the government will, at some point, call for live fire to squash the riots when they rise out of control.

"They will not respond to those requests. Instead, they will release copies of the requests to the public. We figure by the end of the week, it will be at a tipping point. That's when we will need you to go in and eliminate the primary six while we take out the rest of the 21. Alyssa will be tasked as a protector during the mission. Other than yourself, Zack, and Jake, she is the last remaining survivor from your graduating class. She has matured a lot—well, a lot for her. You'd be proud."

"That is great news. I'm happy to hear she's still in the fight."

"You do what you can to get the hell out of there tonight and back to your house in New York. If that fails, find a safe place until I contact you. Do what you can to stay alive. God be with you."

Sam smiled. "I will, and thanks, Laura."

CHAPTER THIRTY-NINE

The drones stopped ripping past his hideout about mid-day. That gave Sam a bad feeling. There was no way they called off the search. He thought about ditching the truck and going on foot, but he would be a sitting duck if they located him.

If they were willing to fire a missile at him last night, there was no telling what they might bring to bear now. The truck contained a few tricks to increase Sam's survivability. He decided his best chance would be to cover as much ground as possible with it while he still could.

Darkness arrived as he started the truck. He took back streets and roads in an effort to not come in contact with any law enforcement. He crossed the New York state line and turned off the road, cutting through the woods.

He flipped on his magnetic field tracker and used the device to scan the sky for anything flying or trying to track him. A large target appeared on the scan right away. With the truck on autopilot, he pulled out a pair of thermal binoculars.

"Shit," he said to himself as he realized what it was: a military-grade hydrogen-powered drone.

Equipped with any number of weapon systems, it had the potential to decimate entire areas. He assumed that was why the drone had not fired on him yet. The pilot must have orders to wait until he was well outside of a populated area. They could concoct any story they'd liked to explain what happens next.

Luckily, his truck wasn't just a grocery-getter. He pressed in a code on the, and a new console menu came up with two menu titles: defense and offense. He went to the defensive menu and flipped on the PPF and anti-missile systems. The PPF charge was at ninety-six percent. Two small multi-barrel cannons flipped out of the cab and spun up. The offensive menu provided nothing to help him take down a drone. Sam's brows scrunched in thought. He would need to improvise.

He grabbed a long-barrel ST-200 spike thrower from a hidden storage cache in the floorboard of the back seat. The weapon was accurate out to sixty feet, twice the distance of his hand-held ST-115. The range was still way too short for what he needed.

If he bypassed the power regulator and went straight from the high-power cell to the accelerator system, he calculated a sixty percent chance it would fire once at full power. That might give him the distance he needed. Otherwise, the accelerator might fry and blow up in his face. He weighed the pros and cons for a moment and realized no other option existed. He started making the change when an alarm on his dash went off.

An incoming missile from the drone closed in on his vehicle, streaking in from his seven o'clock. In the rearview mirror, Sam could see the methane-based propellant creating a brilliant blue tail out from the back of the projectile. His two turrets sent out a stream of hypersonic and subsonic rounds.

Sam made the alteration when he discovered some years back that the military started putting PPFs into their missile systems to give them a higher survivability when fired at locations containing anti-missile systems. Sam found that altering the subsonic rounds with the hypersonic rounds would create a weakness in the lower quality PPFs allowing the subsonic rounds a higher chance to penetrate the field under sustained fire. That should provide a better chance to defeat any incoming missiles fired at the vehicle.

CHAPTER FORTY

J ake monitored the scene from a black ops area within CIA HQ at Langley, Virginia. A city drone had located a suspicious truck heading out of town.

Jake guessed it might be Sam when the truck shut off its lights and went off-road. Jake decided to track it until he made it to whatever location he planned on retreating to.

When he was at his most vulnerable, they would drop the entire missile payload on his head. When the two missile defense pods opened up on the truck, Jake realized the drone had been detected.

"Fire a missile at him. Let's see if he can down a PPF-protected missile with those weapons."

The drone contained four standard infrared high-explosive PPF-protected missiles attached on wing pods and one modified PPF-protected multi-warhead mounted fire-and-forget missile that used laser and optical designation to track to target. That little gem was called "The Pig".

The remote pilot fired off one of the standard IR missiles. It tracked about halfway to the target and was hammered by rounds spewing from the twin turrets on the truck and destroyed.

"How in the hell did he down that missile with hypersonic rounds?"

The truck started accelerating and moving in random directions. They tailed the truck for a bit, watching it climb over ravines, drive through shallow creek beds, and get air over embankments.

"That's one sweet ride. Time to make it his coffin. Do we have satellite lock on his vehicle yet?"

"It is taking some time, sir. He has some sort of signal scrambler preventing a GPS lock on the truck," the drone pilot replied.

"Then place a thermal tag on the truck, and we'll track it until we destroy it or he stops. Is the squad airborne?"

"Yes, sir. They are five miles behind the drone and awaiting further orders."

"OK, keep them at that distance. We shouldn't need them. Fire the rest of the IR missiles. That should overwhelm his defenses."

The pilot fired off the rest of the IR missiles in sequence.

"Goodbye, Sam," said Jake as the missiles tracked to the target.

The turrets fired off streams of superheated rounds that took out two of the missiles as they tracked in. The third appeared as though the missile might reach the target until the truck erupted with dozens of high-intensity flares used to defeat missiles by other aircraft.

The missile tracked off and hit near the truck, launching the vehicle in the air. The 4x4 landed hard, but it kept on using evasive maneuvers as the truck flew over the next ridge. They lost sight of it for a moment; however, the satellite tracking stayed true. The truck popped into sight on the next ridge and followed the spine of the ridge while dodging trees as it went.

"Drop The Pig. That should take care of him."

The pilot fired off the multi-warhead evasive target or MWET missile. It was an antipersonnel air-to-ground missile. The warhead burst just before the truck's turrets opened up and separated into twelve different seeking warheads. The truck responded with flares, a heavy smoke screen, and turret fire. The flares combined with the smoke blinded the warheads long enough to detonate on the ridge just behind the truck. The truck burst through the smoke, unfazed by the missile detonation.

Jake slammed his fists down on the table. "Son of a bitch. Light him up with the velocity cannon!"

"Aye, sir."

The pilot strafed the truck with hundreds of hypervelocity rounds. A PPF field deflected the rounds away as fast as they came in. They ricocheted off in every direction, except for the direction of the truck.

"We don't have anything else to throw at him, sir," the drone pilot replied cautiously.

"Oh, yes we do. Fly the drone straight up his ass!"

The pilot looked at Jake and grinned. "Yes, sir!"

The truck climbed up another ridge and found an old logging road that remained in decent shape and accelerated down it, crashing through or jumping over the occasional log. The pilot flew in at high speed right for the truck. The back window slid down, and Sam hung halfway out the window.

"Fire that cannon while you dive on him!"

The pilot opened fired as he accelerated the drone towards the truck Sam was hanging out of.

Jake leaned in closer to the monitor. "Get him before he jumps out!"

Jake thought Sam was about to leap off the truck. He realized too late Sam was just getting a clear shot with his improvised ST-200.

Sam, half-blind from the stream of rounds bouncing off the PPF, waited until the last moment and fired the spike at the drone as the attack craft loomed in on him. The ST-200 split the barrel wide open from power increase Sam configured on the weapon. He was lucky enough to penetrate the side engine and force the drone down earlier than the pilot anticipated.

The drone bounced off the ground hard and clipped the truck, putting the vehicle into a spin. It slid sideways into a log and rolled the truck. Sam bounced around the inside of the vehicle like a matchstick caught in a tornado until the truck came to rest on its wheels.

Sam climbed through the gear strewn about the cab to the front seat. The dash displayed a warning that the autopilot was damaged. He ran through a couple of checks and pushed the throttle of the truck, and it lurched forward. The truck's night vision was out too. He could see well enough by moonlight to drive without the lights on, so he continued on, assuming he lost his pursuers.

Jake radioed the crew in the chopper keeping its distance during the fight.

"Do you still have him on satellite?"

"Affirmative, tracking one vehicle."

Follow him to wherever he's going and then pounce on

him with everything you've got. Do not hold back. He is not to be underestimated. Unless you identify a body, he is still alive. Got it?"

"Roger that, sir."

Jake pulled up the satellite feed tracking the truck as it continued along the ridge.

The technology mapped the ambient air in the surrounding area and compared that with objects within the same area. The truck was much warmer than the surrounding air and stood out as a yellow blip on the image.

"We'll have no problem tracking him to his destination," said the technician managing the satellite feed.

Sam found his way back to a gravel road and pulled up some GPS data on his location. He deviated from his original path by a few miles before losing his pursuers.

At least, he hoped he lost them. The truck was in no shape to fend off any more drones. One turret showed offline, and the other was almost out of ammunition. Just a few flares remained and the ability to generate smoke was offline.

He pulled into the driveway at his home a couple of hours later and put the truck in the garage. The truck trailed steam out of gashes in the body, and a ticking sound of could be heard emanating from under the vehicle. He couldn't waste time looking it over. He went in the house and secured the perimeter defenses.

Reboot installed three PPF generators designed to protect small military ships upon moving into the home. Those would force anyone attacking him to use heavy ordinance to bust through—or come in on foot and breach the house. Heavy ordinance would level the house. As a precaution, he planned on staying in the basement near the escape tunnel for a couple of days.

He grabbed some water and food and headed for the basement. Once there, he pulled out some explosives, a sniper rifle, an ST-115, and secured his sword.

Sam set the explosives in the ceiling near the house's end of the mile-long tunnel. He figured collapsing a two-hundred-foot section would be sufficient. He would use the sniper rifle to monitor the entry of the tunnel from the other end. Everything else he grabbed was for survival or close-quarters combat.

He made his way to the safehouse at the far end of the tunnel and pulled up the security cameras just in time to see a carrier drone coming in from the south. There was no place for the drone to land. He assumed they would either blow up a place to land, fire on the house, or repel down and come in on foot.

A salvo of rockets pelted the driveway leading up the house, leveling an area big enough for the chopper to land. At the same time, a burst of hypervelocity rounds streamed in towards the house. They ricocheted off the home's PPF and put on a brilliant light show for anyone within miles to enjoy.

He watched two missiles leave the chopper, and a salvo of rockets headed straight for the house. He lost his feed as they hit. A couple of seconds later, the shockwave came through like a clap of thunder. Even the feeds in the basement were not accessible. He left the monitor and went back to the end of the tunnel, shutting down the lights in the tunnel with a touch of his hand to the wall plate. He flipped a red flashlight pen on, found his detonator, and readied it. He closed his eyes and whispered, "This isn't their first rodeo. Let's hope it's their last."

He turned off the flashlight and flipped the protective cover up on the scope of the rifle, turning on its heat-mapping feature. He could make out the outline of the hallway and the fading heat from the lights along the walls.

He left the door closed and barred it from his side. The door was thick and explosive-resistant, but it wasn't impenetrable. If they didn't turn the house into a crater, then they would attempt to identify his body. If they were halfway competent, they would find the door and get past it.

He spread out on the floor in a prone firing position for twenty minutes, peering through the scope of the rifle. He was about to give up, blow the tunnel, and escape through the safehouse above him when the door at the far end erupted in a halo of fire. Instead of trying to blast through the door, they cut around the door frame where it was weakest with what appeared to be thermite plasma.

Thermite was an intensive pyrotechnic that could melt through anything given enough time and fuel. The whole door and frame fell forward with a heavy thud, kicking up smoke and dust.

Sam clicked the magnifier up a notch to get a closer look at the doorway. The detonator lay just a few inches away from him. The smoke settled, and he saw no movement for a couple of minutes. Then, a small device peered around the corner. Sam held his fire. He knew a device of that size would not be able to make him out over a mile away in total darkness. Even with heat mapping, the scanner would not have the required resolution to spot him or to sense heat from his location.

What he didn't expect were the two autonomous hounds that bounded through the doorway. They came forward at a trot. He held his fire, hoping a human might enter the hallway with them. He changed his tune when the hounds switched to a sprint—they'd caught his scent. He fired at the lead hound. It ricocheted away from the hound before impact.

"Shit. Assholes put PPFs on their hounds."

Sam slapped the detonator, and the far end of the hallway exploded. The speed the hounds were traveling took them past the blast zone. They were about half a mile out and gaining fast. He fired a few more rounds at them, but it was no use. He needed to lock them in the tunnel or take them on at close range.

He didn't want to take a chance of close-range combat with hounds again. They maintained the advantage in close quarters such as these. However, the team would be able to use them to track him, even in shadow.

Sam decided it was better to stand and fight now. If he took out the hounds, he could take on the team on his home turf. If he ran, he would be at their mercy.

Sam stood up and kicked the rifle to the side. He slapped the wall with his hand, and enough of the wall lights lit up that he could see the hounds remained about a quarter mile out. Their matte black metal bodies resembled two torpedoes homing on him.

He brandished the ST-115 in his left hand and pulled the molecular sword from its sheath with his right. The stake thrower held just two shots. If the hounds mimicked the ones who came for him at the White House, they would barrel straight for him.

As they approached, he let loose both stakes at the hound to the right. The first stake penetrated the PPF and embedded itself in the hound's shoulder. The hound lost its footing just as the second stake found the opening in its mouth. The stake disappeared into the back of the hound's throat, hitting something important and disabling the mechanical beast.

Sam pushed off the wall to his left and leaped over the hound he shot. As it slid past, the second hound dove at him, its metal jaws snapping shut on thin air as he brought his sword down on the hound's back. Both halves slid past him and slammed into the door behind him.

Sam picked up the rifle and peered back down the hallway. The tunnel was blocked by the cave-in. Nobody was coming at him from that direction. That gave him a few minutes to examine the hounds.

The first hound was still flopping around until he cut the head off. That meant the processor was in the head. The jaws were hydraulic on this version as well. That would give them a very strong bite. By the looks of it, they had enough clamping force to sever the limb of an average human.

Hydraulics made the hounds heavy. That didn't seem to slow them down, but it hindered their agility a great deal. Their metal claws gave them excellent grip on soft surfaces. The claws remained a detriment on a hard surface such as the tile floor Sam stood on.

Sam examined the hound he cut into two pieces. The hydraulic pump for the jaw was located in the chest. One high power output cell sat where the anus would be on a live specimen.

The tail was whip-like and smooth. The leather-like tail mimicked the tail of a real wolf or dog, keeping them balanced. The rest of the animal was powered by electric servos.

The technology and their weight made them useless in water. Sam could tell these were prototypes too. The two, though similar, contained slight differences in their make and build. They were built by hand.

Sam snatched up the head of one of the dismembered hounds and took it with him. Leaving the corridor, he went up into the safehouse and stashed the head in a bag he took with him when he left.

He grabbed a communications scrambler similar to what he'd used in the limousine. The scrambler would block their ability to communicate with the receiver in the drone circling above.

He shifted some things around in the small area to unlock a panel in the wall. Once unlocked, he opened it to reveal a ground-to-air launcher. He hoped he would never need it, but he was happy to have access to one now.

Sam slung the launcher over his shoulder and squeezed through the trap door opening that led to the woods. After making his way outside, he took out the comm scrambler, hid the bag, and turned back toward the house about a mile away. Time for some pay back.

He slipped on his magnetic field visor and went into shadow. He had no further rounds on the stake thrower, so he tossed it aside. He checked the charge on his sword and took off in a sprint towards the house.

The operators who hit the house were making their way back out through the wreckage. They tracked the hounds with GPS and watched them close in on Sam via video stream. After contact, there was nothing but static.

The leader of the five-man team was a long-time veteran of the Navy Seals, John Redcrane. He was part Native American and had enlisted in the military to escape the poverty-stricken reservation life. He retired after twenty-two years and turned to contracting when civilian life didn't suit him.

The other four members of his team had been together for a few years. They functioned like a well-oiled machine.

When asked why the Secret Service didn't head up the man hunt, they informed him they might have a mole. He didn't know if the information was true or not, but the money was good, and in this climate, that was all that mattered.

After watching Sam dispatch the hounds with relative ease, John realized this was no ordinary man. The cave-in left no way to pursue the mark.

The hounds were about a mile out. He scanned the area with the chopper, circling above the scene. They found something interesting near the area where they lost the hounds—some sort of capsule or structure hidden by foliage the scans picked up.

While watching footage of the structure, a figure came out of the capsule's hatch and headed back their direction. He told the pilot of the drone to track the target and keep them updated on his location. He waved everyone back out of the basement. He spoke to the rest of the team when they got topside.

"Guys, our mark is headed back this way after dispatching the hounds and escaping through a structure at the far end."

"That doesn't make sense, Crane," said his second lead, Joe Tucker.

Joe was a smart tactician and a solid second in command. They had a lot in common. Neither could deal with civilian life, and both had been Navy Seals for decades.

"No, it doesn't. But the fact remains he is. Halo is going to keep us posted on his location. We're going to spread out in a five-meter gap line formation and locate the target. Kill on sight is still the order."

Everyone nodded in confirmation. The group streamed out of the burning house, left half-standing after the strike from the chopper. They headed into the woods in the direction Sam was coming at them from.

Halo relayed to John, "Target moving at around twelve miles an hour at your one o'clock. His position is incoming at four hundred and fifty yards and closing."

"Roger that, Halo," Redcrane responded. "Go ahead and light him up."

Redcrane motioned to the others to proceed to the top of ridge where they dug in.

"Switch to infrared tracking," Redcrane said through the comms.

A moment later, they watched the drone let loose with its cannon, firing hypersonic rounds at Sam. A light show of reflected rounds flew up from Sam's position as the PPF field deflected the shots away from Sam.

After the drone ran out of rounds, the team watched a missile streak out towards the drone a few ridges away, striking the drone dead center. Redcrane and the team watched in surprise as the drone came crashing into the woods a few hundred yards away.

"There goes our ride," commented Redcrane. "Keep your wits about you. This target knows what he's doing."

Sam ran through the woods as fast as his legs managed while dodging trees and other undergrowth. He was getting close to the house, and they knew he was coming.

He low-crawled to the top of the ridge and set his visor to record. He pulled the slim-faced visor off and held it just above the ridge, panning the device from left to right. He then slipped it back on and replayed the recording.

At the top of the adjacent ridge, he distinguished five magnetic signatures all along the ridge set for an ambush. His familiarity with the woods would allow him to flank them now that the drone wasn't around to track him. From this moment on, they would be on their own, and he would have the advantage.

Sam pulled out the comm scrambler, turned it on, and left it there. Staying low to the ground, he followed the ridge and worked on flanking to their left where their ridge led down to a creek bed.

After a few minutes, Redcrane started getting a bad feeling. He squinted through his infrared lens.

"Anyone see movement?"

No response.

"Comms check."

None of the team heard him.

He flipped on his wrist pad and tried pulling up video streams from any of the team members. No connection was available. He realized they were being jammed.

"Shit ..."

Redcrane motioned to the rest of the team using hand signals to indicate comms were out and to watch for flanking maneuvers to the left or right. They confirmed his hand signals and made appropriate adjustments.

Mike Tragen, their munitions and explosives specialist, was near the far-left flank. Craig Sticklman, their heavy weapons expert, was on the other side of a large tree about fifteen feet to his left. Tragen heard something hit the tree with a thud as Sticklman's Gatling spike gun hit the ground and rolled to the bottom of the ravine.

Tragen shifted his position to get a better view of what was going on and found Sticklman pinned to the tree with a throwing knife through the neck. He scanned around while back-peddling up the ravine towards the remaining team.

"Sticklman is down!" he cried out.

"Zero formation!" Redcrane shouted.

The four remaining men shifted into a circle formation with weapons out, looking for Sam.

"Does anyone see anything?" Redcrane asked.

"No ... wait. Our nine o'clock!" called out Croomile, their melee specialist.

Before anyone could turn to that direction, Sam appeared in the middle of their formation.

"Looking for me, gentlemen?"

Before they could react, he kicked Redcrane in the back, sending him face-first into a nearby tree and knocking him unconscious. At the same time, he reached out and thrust his sword through Tragen's armored backplate. The tip burst from his chest, and with a twist, Sam yanked it back out. Tragen collapsed on the ground, his face twisted in a silent scream.

Sam turned with a spinning kick, catching Croomile on the chin and sending him sprawling. Tucker whirled around with his spike gun and fired off a few rounds as Sam caught the barrel of the weapon in his left hand. The rounds flew off into the woods, making thudding sounds as they embedded into trees.

Tucker threw a right cross that connected with Sam's chin. Sam spun with the blow and brought his sword around, taking off Tucker's head. The head rolled down the hill, splashing into the creek at the bottom.

Sam heard a scream and turned to see Croomile storming back up the ridge at him, brandishing two swords. Sam readied himself as Croomile came in with swipe after swipe. Sam parried each but received a kick to the chest from Croomile. Croomile kept the pressure on with savage blows. Every blow, Sam parried away.

Croomile was screaming the entire time. He'd lost control and wanted to chop Sam into little pieces. With the constant blows being raining down upon him from Croomile, Sam couldn't find an opportunity to strike back.

After a minute or two, Sam could tell Croomile was losing his strength from the barrage of strikes he attempted to land.

On Croomile's next strike, Sam turned his blade so that it hit edge to edge on Croomile's sword. The molecular edge of Sam's sword cut Croomile's blade in two. The broken blade's tip went flying, embedding itself into the tree Sam had kicked Redcrane into. Croomile never missed a beat, switching to two hands on his remaining sword.

Sam twirled away and crouched down into a low sweeping kick, taking out Croomile's legs, who landed hard on his back. Sam allowed him to scramble to his feet. Both men stood some distance apart, glaring at each other.

Croomile panted as spittle flecked his face and dripped off his chin. He was almost foaming at the mouth he was in such a rage. Sam understood all too well how Croomile was feeling. He just watched his entire team get taken down in a matter of moments.

They left Sam little choice. Sam learned long ago that giving in to rage would always end in defeat. It sapped a person's energy and clouded their mind. It pained him to do it, but he gave Croomile a little grin and wink.

That was all the instigation Croomile needed to be sent into another fit of rage. He came at Sam with a hard slash from left to right. Sam rolled under the slash and severed Croomile's leg in the process.

Croomile fell on his back. Sam came out of the roll, jumped up into the air, and came down with a heavy blow that Croomile tried to parry with his own sword. As before, Sam's molecular sword cut through the second blade and deep into his chest. Blood splattered across Sam's face as he cut into Croomile's heart. The brave man died seconds later.

Sam turned to Redcrane, who remained unconscious after his encounter with the tree. Sam knew he should finish him off, but he couldn't kill an unarmed man.

The decision may cost him later. It was hard enough living with the perfect memory of every person he ever killed. He turned away and retrieved his comm scrambler and bag from the woods. It had a five-mile range on it. That gave him some time to disappear.

Sam slipped on an IR poncho similar to the one he used when he first escaped from D.C. out of the bag he'd pulled from the safehouse. He was effectively invisible to IR scans with it on. He headed off towards a safe house he had picked out when he first moved in as a precaution.

CHAPTER FORTY-ONE

S am made it to safety and spent the next couple of days locating his backup vehicle and getting to another safehouse Reboot had listed for him in case his identity was discovered.

Somehow, Jake had figured out who he was, and Sam didn't know what had given him away. Maybe Jake just wanted Phil removed from the equation.

Regardless, Sam had managed to escape. Once he was able, he contacted Laura, who confirmed all major cities in the U.S. were having protests or riots, as planned.

The military had been given the go-ahead to fire live rounds at protestors and rioters. They ignored the order and continued to use non-lethal methods to manage the crowds. Reboot leaked the order to the public, which led to a movement by the people en-mass against the government.

The situation was something Reboot did not expect but remained grateful for. Citizens marched into the nation's capital. Tens of thousands of protesters took over the landmarks with a sit-in that proved difficult for authorities to break up. All the local jails were at capacity, so a special location was being erected to house additional arrests until another riot broke out and it was destroyed.

People called for a reconstituted government with proper representation and no more lobbyists. Reboot jumped on the opportunity, delivering video screens and other communication equipment to locations for protesters.

Reboot presented a two-hour video loop of all the atrocities the government tried to hide ever since Reboot presented the world with a cure for communicable diseases years before. All known evidence of pressure, bribes, and even assassination attempts were presented by Reboot to the people.

Reboot took the lead and gave the government officials an ultimatum. A list of all officials, which evidence showed committed crimes against the citizens, were to be removed from office and charged for those crimes. A list of officials who were coerced and remained in good standing were to be allowed to stay in office until the new election process was instituted.

During the video being presented in Washington D.C., the Group of 21 sent every agent, police, and security force to surround the government buildings and shoot anyone who attempted to enter the locations. This started a riot, which soon turned bloody when both sides began using explosions and live rounds.

Reboot didn't advertise their presence. They continued to work from the dark, but every clip of evidence came from their research division.

Laura contacted Sam and gave him the go-ahead to meet with Alyssa and formulate a plan to end and remove the 21 by any means necessary. A division of the Marines under the command of the retired general they'd contacted was marching on the city with a few dozen technicals, a non-armored improvised military vehicle, and two stolen C_1Z Battle Tanks.

The move by the military was a distraction Reboot hoped would be enough to get company operatives within the capital building to surrender, along with other primary government facilities.

The six leaders of the 21 were to be eliminated for their crimes. This prevented them from using their contacts with lobbyists and government entities to stay in power.

Sam contacted Alyssa and the general. Sam was to meet with them at the general's forward operating base outside the city limits at the Ronald Reagan Washington national airport. Alyssa confirmed her time of arrival to the FOB, as did Sam.

CHAPTER FORTY-TWO

S am drove up to the gate of the base. The guard scanned him but didn't find a match in the system. For a moment, it looked as if the guards were not going to let him in until Alyssa showed up.

"Look what the cat dragged in!" Alyssa said as she walked up to the gate guards.

"I'm not certain if we want to let in your sort."

"My sort? What sort is that?" Sam said with feigned surprise.

"Oh, you know the kind ... all business, doesn't understand the meaning of fun." Alyssa smiled with glee.

"That's a low blow. I know how to have fun. I've been a bit busy, you know."

"Oh, really? You think changing your face and being on the run counts?" She laughed. "He's good, boys. Let him in."

Sam smiled for the first time in a long time. After everything she must have been through, Alyssa still retained her sense of humor. She was a breath of fresh air for Sam.

Society was on the brink of a major change in power

back to the people. Sam hoped it was a change for the better. They couldn't trip at the finish line. Thousands more would die if they did.

The guards let Sam drive through the gate. Alyssa got in, and she showed him where to park. Once they both got out of his car, Alyssa came around and almost leaped into his arms for a long hug.

She gave him a peck on the lips.

"You still smell the same."

"Some things never change, and some things do." He gave her another squeeze before they started walking towards an aircraft hangar.

"Well?"

"Well, what?" she asked.

"What do you think?"

"About what?"

"About my face?"

"Ohhh, your face? Since you asked—I prefer the original. But this one will do for now."

Sam chuckled. "Not the best response, but I'll take it."

"I've got a surprise for you before we meet with the general."

"Is this a good surprise or one of those ones I could do without?"

"You and your glass half-empty attitude, I swear." She jabbed him in the ribs. "Of course, it is a good surprise. I'm part of the surprise!"

Sam laughed and paused to stare at her. "It's great to see you. I realize now how much I missed being around people I can talk to."

Her face turned serious, and her eyes grew misty for a moment. She blinked a couple of times and regained control. "We'll talk more later. First, your surprise."

She assisted Sam in getting issued a temporary ID. Then, she escorted him through a bunch of military members running around between desks and monitors to the back of the hangar that had been cordoned off from the rest of the building.

In the room, a large holographic monitor showed a three-dimensional image of the Washington D.C. area.

"This is the final stage."

"Is this my surprise?"

"No, no ... She is."

Alyssa pointed past Sam to a woman who entered the room. Sam turned to find Laura Cinder standing there smiling at him from ear to ear.

"Holy shit!" he exclaimed.

He walked over to her and gave her a big hug lifting her off the ground and spinning her around.

"What in the hell are you doing above ground?"

"This is it. This is what all the blood, sweat, and tears helped us get to. Everyone here wants America to be the United States of America again. A world power, yes, but a world power who uses their power to help the world recover and become one. Not one who uses it to coerce the rest of the world into doing what it wants. Here, I don't need to hide."

"This is great! It's been a long time since I have been this happy," said Sam. "What's next?"

As if on cue, the General walked in with an assistant.

"Sam, I want to introduce you to General Forrest. His ties to this country go all the way back to before the Civil War. He is named after Nathan Bedford Forrest who is his ancestor going back many, many generations."

Sam looked at him. "Pleasure to meet you, General Forrest. I happen to know a bit about General Nathan

Bedford Forrest, the Wizard of the Saddle. He stood out as one of the Confederate's best generals in the American Civil War."

The General raised his white bushy eyebrows and glanced at Laura, then back to Sam. "How on Earth do you know anything about my family history?"

"I like to study anything I can get my hands on, sir. I like to think history is something we should all have a better understanding and respect for. Plus, I did a bit of background on some of the key contacts for us. Your last name, for example, is not Forrest. Forrest is your middle name, and Nathan is your last name.

"The Forrest name died out a few generations after Nathan Bedford Forrest, but your bloodline continued to be passed on via the women in the family. History is one of those things we should all learn from. That way, we can avoid making mistakes we've made before or recognize them when they start to happen."

Laura interrupted. "Sam is being a bit humble, General. He has a genius-level IQ."

The General grinned as he walked past Sam and Laura to the hologram map. He was a commanding figure even at his eighty-plus years of age. "Miss Cinder, you know how to pick 'em."

"Thank you, General."

The General stood staring at the map for a moment. The rest gathered around.

"Our forces are going to surround the D.C. area and start moving to the Capital building. Along the way, we will subdue any forces who resist and corral the rioters. Once we push forward to the Capital building, we will secure the roads and major monuments around the area. There you can get your message across about the changes planned for the future. I assume someone from your staff has been selected to give a statement?"

Laura nodded. "Yes, our company historian is going to present the announcement, call for a cease of resistance from both rioters and police, and give an explanation as to what the next steps are."

"And what of the six gentlemen you spoke of?"

"They are being executed for treason," replied Laura.

"What if they give up?"

"They won't," Sam said.

"What makes you think that?"

"Sir, they've killed, coerced, blackmailed, bribed thousands we know of, and affected millions with their schemes. They know what's in store for them if they give up."

The General scratched his bearded chin. "I suppose our country was founded on blood. I will not object to any action you take. You're the one who must live with your decision, not I. This division is here to secure the city, not occupy it."

"I have not been sent as their executioner, General. They have been judged by someone else."

The General stared at Sam for a long moment. "I can see you have a personal stake in this. Be careful of the line you walk. Don't let them pull you into an emotional decision and take away your humanity. Anyone you lost cannot be brought back, son. Killing in their name doesn't make the pain go away."

Sam waved away his comment.

"No, I understand. I have no intention of taking revenge. It is not my place to judge or condemn."

"I hope so."

The General turned to face Laura.

"Well, Miss Cinder, let's get the show on the road. We'll

position ourselves around the city and start pushing in at three o'clock tomorrow afternoon. That should give you enough of a distraction to allow your agents to infiltrate the target locations and complete your missions. We will pop flares once the Capital building is reached. Then, it will be up to you."

He walked over to Alyssa. "You, my dear, have been a Godsend. Good luck in your next foray into the unknown."

Alyssa smiled. "Thanks, General." She hugged him. "It was an honor and a privilege."

CHAPTER FORTY-THREE

The meeting broke up, and everyone went off to prepare for the final push the next afternoon. Alyssa made certain to brush past Sam and gave his butt a squeeze as they went off to gather their gear. Sam smiled from ear to ear. Same aggressive woman he knew from training.

He got his gear from his car and was given a bunk in a makeshift barracks. He took his time checking over everything he planned on bringing for his hunt of the six congressmen and Jake.

His sword, tactical gear, ST-115 with additional rounds, a concussion grenade, and an automated entry sealer composed the majority of his gear.

The door sealer was the same type he'd used to prevent the Secret Service from entering the President's chambers. He was changing the battery out of the sword's hilt when a knock came at the door.

"Come in."

Alyssa opened the door. Her black hair hung to just

below her ears. The tips dyed a fresh dark blue and faded into what Sam guessed was close to her natural hair color.

His heart rate increased at the sight of her.

She had changed into some form-fitting charcoal grey pants with a canvas belt and a dark blue crop top that matched the tips of her hair, showing off her toned stomach. She loved showing off her body.

Sam glanced up and then back down to his sword. He paused, then looked back up at her, taking her in. He felt the same lump in his throat as he did when they'd met after graduation. That seemed like ages ago.

She stood smiling. He realized he was staring and looked back down at his sword.

"You'd think with your perfect memory, you wouldn't have to stare so hard," she said, closing the door behind her.

"Beauty is a rarity in this world. It's best to appreciate it when it arrives at your door."

Sam could tell he was blushing. She strolled over and sat on his cot, crossing her legs.

"You always know what to say to make me swoon. I assume it works on all the other ladies as well?"

Sam shrugged. "What other ladies? I've been in deep cover or on the run since graduation."

"You're telling me that you didn't bump into any women since me?"

"No ... Plus, we were never in a relationship. You made your point clear that you didn't want to start anything."

Alyssa fiddled with the drawstring on Sam's pack he was filling with gear.

"For good reason, you know that. You were headed off to God knows where, same for me. I didn't know if we would ever see each other again. We both lost friends and made enemies along the way."

Sam stopped fidgeting with his sword and looked her in the eye. She was right. He could have died any number of times between the last time they spoke and now.

"You're right, but that doesn't mean I don't regret it. In the last week, I've almost been killed three times. I was prepared to pay that cost. The one regret is that I couldn't hold you, take in your scent, or kiss you one last time."

Sam walked around the end of the cot as Alyssa stood, pushing herself into his arms. She slid her hand up through his short hair, smiling softly. It was as if he were in a time warp. She smelled the same, and her body felt the same pressed up against his. She whispered into his ear.

"If all goes well tomorrow, I'll be giving you a workout."

Sam pulled away a bit to look into her eyes. He ran his hand along her bare waist before pulling her back to him. Their lips touched and parted as they tasted each other for a moment. Sam smoothed a hand over her hair. They finished with a long hug as Sam whispered back.

"I'll hold you to that."

She pulled away and smiled at him.

"Come on, let's go enjoy some dinner while you tell me about all the craziness you've been into since graduation. I've heard the rumors, but I'd like to hear it from the source."

"As you wish."

Sam slung his sword on his back.

"You carry that thing everywhere. Why?"

"Next to you, this sword is my most dependable friend. It saved my life many times."

"Maybe you can put it away after this?"

Sam nodded at her but said nothing. Alyssa noticed the expression on Sam's face. The look reminded her of her father when she'd asked him if his tour was the last one he was going to go on, until the one where he never came back.

She kept her thoughts to herself as they walked out the door.

The two spent the evening enjoying each other's company. They reminisced about when they first met, held each other, talked about the mission, and snuggled together before falling asleep in each other's arms.

The next afternoon, the two hugged each other before heading off in different directions. Alyssa gave him a deep kiss before she grabbed her gear. Turning, she locked eyes with him.

"Good luck, sexy. I better see you later."

Without allowing him to answer, she turned and walked off. Sam watched her walk away, counting the seconds in his head. At fourteen seconds, she looked back at him.

When Sam gave her a little wave and a smile, she realized he was waiting for her to turn like she did after graduation. Her face turned pink. She looked away, smiled, and glanced back at him. Sam softly laughed to himself and turned. If he were lucky, he would be able to fulfill that request of hers.

CHAPTER FORTY-FOUR

His mind turned to his mission. He knew his target. Surveillance Drones indicated a majority of the congressmen and women gathered at the Capital building. That was ground zero for Sam.

As the convoy moved through the streets, they met little to no resistance. The military moved in and started dispersing the protesters away from the target location. Military drones flew overhead. The few drones that came out to meet them were quickly dispatched. The General's division maintained air superiority and allowed a much easier time traveling than Sam expected.

Sam followed along with the military forces for a while, dressed in their uniform to blend in. They progressed to within a block of the Capitol. As the military approached, a sudden rush of drones came out of hiding, strafing the front lines of the military formation with smoke grenades.

The drones continued to fire off smoke grenade rounds all around the perimeter of the Capital Building. This continued until Sam couldn't discern objects more than fifteen feet in front of him.

Soon, military drones came in and started thinning out the security drones responsible for the smoke. A few casualties were reported. Smoke grenades continued going off after the drones were downed.

Based on trajectories and lack of drones, Sam could not figure out where the grenades were coming from. He was about to leave the team leader when a message came across the public channels and intercoms around the city.

"This is your government—we are in charge. If you come any closer, we will start executing cabinet members, followed by congressional members. You will disperse within fifteen minutes. Go back to your bases and disarm."

Sam looked over at the squad leader. "Do they think that would work?"

Sam didn't know how serious they were about the accusation. He decided the time came to proceed on his own.

"Sergeant, I'm headed off." The Sergeant and Sam thumped their forearms together as a sign of respect.

"Roger that, sir. Our orders are to standby."

"Understood, sergeant."

Sam took off the military uniform and proceeded forward in his tactical gear. The smoke hindered his sight, but it did allow him to move forward without being seen. He heard the pop of flare guns being fired off in the distance and the faint glow of pink seen through the dense smoke. The convoy was in place.

When he found the steps of the Capital building, Sam realized the reason for the smoke. Dozens of autonomous hounds could be heard roaming the perimeter. The clink of their metal claws on the stone and concrete echoed all around him. He readied his sword and ST-115, doing his best to hide in shadow.

He sprinted forward until he found the pillars of the

main entrance. Just as he did, the familiar gait of a hound sprinting towards him reached his ears.

Sam tried to prepare himself for the hound. Instead of one, two came out of the haze of smoke. He parried the lunge of the first with his sword. But, before he trained his ST-115 on the other hound, it slammed into him. Its jaws snapped inches from his face and knocked his ST-115 out of his hand. He kicked the metal beast away as the other came back at him. That one lunged at him, and his sword went right down into its mouth. Sam slashed down, cutting the sword free. He followed that with a kick to its face.

He rolled forward, grabbing his ST-115 as the other hound shot past him. Its jaws clamped down inches from where his arm had been a second before. The hound skidded down the steps and into the smoke. Sam used the opportunity to use his holy strength and sprint for the main entry doors.

He was able to give himself a good lead as he bounded for the double door. The hounds tracked by scent, and he heard more on the way. As he made his way to the front doors, he ran up on a security agent who was bent over, reloading a grenade launcher. Sam ran him through with his sword before the man could react.

Sam dragged him inside, along with the bag of grenades and the launcher. Yanking the entry sealer out of his back-pack, he slapped it against the top of the entry.

The mechanism unfolded into place just before he heard a hound slam into the door. The door held as more and more hounds slammed into it. He heard their metal claws scratching at the wood, attempting to get through to him.

They wouldn't be getting through this way. He ignored

the scratching sounds from the door and surveyed his surroundings.

No other security could be seen. Nobody was around, for that matter. He loaded the twelve-round grenade launcher with smoke grenades, firing them into the hallways and offices as he progressed deeper into the Capitol.

During Sam's stint protecting Jake, he studied the floor plan of the building in detail. The Capitol building rose three stories high and contained a basement. The House and Senate chambers were located on the second and third floors. The House chambers resided on the far south end of the building, and the Senate chambers on the far north end of the building. Sam did not know where the six men might hold up in the building, but he doubted he would find them in the House chamber. However, he needed to secure any congressional members not in the remaining Group of 20.

Gorden Linden was no longer considered a conspirator. Alyssa's mission required her to secure him prior to him arriving at the Capitol.

That left 20 conspirators, and of those, the six that led them. Sam knew they would try something convoluted. He didn't expect them to kill their own, but people do crazy things out of desperation.

Sam proceeded down the main corridor. The building felt empty. His steps echoed off the walls and down the hallway. No other guards patrolled this area, giving the building an eerie quality.

When he heard the clicking on the marble floor, he understood why. A hound must have been alerted by the noise he'd made shooting off the smoke grenades.

The clicking sound came from the south side of the crypt, which was the central dome of the building that led to the north and south wings.

Sam decided he would cut to the chase. He dropped the grenade launcher and ran the rest of the way down the hall to the crypt. Just as he rounded the corner, he came to an immediate halt. Coming out of the south corridor was the biggest hound he encountered so far.

"Oh shit," he said, tensing.

The beast locked eyes with him. Its shoulder came up to Sam's waist, and it reached six feet long before the tail started. Sam's brain went into overdrive as he thought about what to do next, looking around him for anything that could help.

He whipped out his ST-115, firing off both rounds in quick succession. The hound caught the first stake in its mouth, twisting slightly so that the second spike landed in its left shoulder.

The hound opened its mouth and let the stake drop to the floor with a loud ringing sound as it skittered along the floor. If it had been trying to determine if Sam was hostile or not, the stake sticking out of its shoulder squashed any doubt. The hound crouched down and launched at Sam.

Sam turned and ran towards the far wall. An idea came to him as he sprinted at the wall, the hound closing in with every stride.

Sam reached the wall and launched himself at it. He landed about four feet up on the wall, using it as leverage for the footing he needed. He shot high up into a back flip and caught sight of his target.

The hound slammed into the bench just below where Sam leapt off the wall. Sam came out of his flip with his sword slicing right through the back hindquarters of the hound, splitting them in two.

He leapt backwards and readied himself for whatever

fight the hound had left. The hound's back legs were immobilized.

The hound clambered after Sam using its front legs as crutches. Its dagger-like teeth gnashed down as the hound's black eyes locked on to his every movement.

Sam dodged to the left and right, using mobility to his advantage. Whenever the hound got close, Sam slashed at its face. After a few strikes, the lumbering beast changed strategies.

It dragged itself in a circle around Sam. Before Sam could react, the beast swung its whole body in a sweeping circle that Sam evaded. It was the tail that Sam misjudged. The leather-like tail took his legs out from under him, and the beast leapt on top of Sam.

Sam was able to get his powerful legs between the metallic animal and him and shoved it away with all his might. The hound lashed out with its claws, catching Sam's thigh as it slid away on the slick marble floor.

Sam didn't have time to worry about his leg as he scrambled to his feet. Remembering what the hound did with the stake he fired at it earlier, he threw his ST-115 at the hound's face and followed right behind it with a slash of his sword.

The hound caught the weapon, crushing it to bits as Sam's sword sliced its face in half. The hound slumped to the floor. Sam cut its head off for good measure and retrieved what was left of his weapon.

"That was my favorite one," he muttered, kicking the massive hound's unresponsive body.

He sheathed his sword and headed for the Hall of Columns. As the name suggested, the hall was lined with fourteen pairs of columns made of marble.

The marble floors where white with black marble

accents. Sam passed through a side hallway leading to stairs on the west side of the building. He used those to ascend to the second floor where the House chambers were located.

As he neared the second floor, he found two guards talking. He moved up the stairs as fast as his legs would go.

The two surprised guards didn't have time to raise an alarm before Sam had run one of them through and throat-punched the other. While the second guard was choking, he stabbed him through the heart.

He decided to keep climbing up to the third floor that overlooked the House chambers. He reached the top and snatched the first congressman he came to, dragging him back to the stairs.

"Don't scream. I've been sent in to save you. Don't yell out, OK?"

The man nodded. Sam removed his hand from the man's mouth.

"How many members are here?" asked Sam.

"Maybe two hundred total," the man whispered back.

"Has anyone been hurt?"

"They let some of those beasts maul of few of the members as an example of what would happen should any of us get the idea of running. Others were ruffed up by the security team. Other than that, nobody is seriously injured."

Sam gave a puzzled look.

"Why are there not more security?"

"I think most of them ran off when the military moved in."

"Then why haven't you overwhelmed them and gotten away?"

"We would have, but they have those mechanical robot dogs that they used to round us up. Have you seen those things? Ghastly looking creatures."

Sam patted the man's shoulder.

"I suppose I would cooperate too."

"They also said they planted enough explosives around the building to level the place to the ground. If they saw any of us leaving, they would bring the entire building down on top of us."

Sam glanced around. "Hmmm. I doubt they have explosives rigged on the entire building."

"How do you know?"

"Bad people still have a survival instinct. If they rigged the building, you can be certain they're going to be in a safe place when they blow it up."

"A few of us saw the lot of them heading to the north wing while they were rounding us up."

Sam nodded. "That's what I expected. They must have holed themselves up in the Senate chambers. OK, stay here and spread the word. Myself or someone like me will come and take you to safety. For now, we need to play along. I'm going to go find the trigger for the explosives and disable it."

"OK."

Sam let the congressman go back into the House chambers, and he sprinted down the stairs back to the first floor to look for a support column. that's where the explosives would be planted.

Sam slinked back to the first floor and located a support column. The congressman was correct—the bastards rigged the building with shape charges. An idea came to him as he examined the rigging.

He disarmed and cut down a piece of one of the shape charges. The shape charge was designed to explode in a specific direction. This left little damage outside of a specific cone while anything inside that cone was obliterated.

He didn't have time to find all the primary charges. They could also have secondary charges planted around the building he would not be able to locate. He wired the shape charge, then used some parts from his damaged ST-115 to wire up a makeshift detonator.

Sam snaked his way from the south wing to the north wing by means of secondary hallways and rooms. He bypassed a couple of security guards on patrol and one hound that he heard from a distance.

He progressed through to the first floor of the Senate wing and found the stairs to the second and third floors.

As he moved to the exit on the second floor, a security guard got the jump on him and pressed the barrel of a gun to the back of his head.

"OK, buddy. Don't move. I doubt a PPF is going to save you from a point-blank shot."

Sam held his hands up for a moment, analyzing them man's stance. A heartbeat later, Sam slid down and back in one fluid movement firing his elbow back into the armpit of the security guard. The shot fired off and went into the floor.

Sam flipped the guard onto his hands and knees and grabbed for his head. Just as he wrapped his hands around the guard's head, two hounds came in through the second-floor entry, followed by Jake, brandishing both of his molecular swords.

"Ah, ah, ah. Let the guard go, Sam."

Sam looked up at Jake and the dogs. "Oh, hi, Jake. I was looking for you." He let go of the guard and stepped back.

The guard gathered his hat, weapon, and Sam's sword. He reached back and slugged Sam in the face. Sam laughed.

"I suppose I deserved that."

Jake raised both swords and sighed. "Do you know how

much of a pain in my ass you've been? Now come this way before I have these beasties tear you apart. You can't take all three of us, special skills or not."

"Alright, alright," said Sam. He raised his arms.

Sam walked up the stairs to the third floor with both hounds flanking him and the tip of Jake's sword digging into his back. He could feel a trickle of blood running down to his waist. The guard he almost killed led the way.

They came to the third floor, took a left, and then an immediate right. That led to a corridor that flanked the balcony of the upper Senate chamber. Two security guards joined them at the west stairwell. Just past the stairwell, they pushed Sam into a large room.

"Gentlemen, may I introduce the pain in your ass, Sam Michelson Creedy."

In front of Sam sat Senators Bryan Lambert from Texas, Jeff Patterson from California, Scott Browling from New York, Louis Mantel from Florida, Jamie Christianson from Texas, and the Speaker of the House, Gregory Crocker.

Sam surveyed the room and identified what looked to be a detonator sitting on a desk in front of Senator Lambert.

The two hounds positioned themselves between Sam and the congressmen. Jake stood back and to Sam's right, sheathing his swords. Next to the detonator in front of Lambert, the two guards laid the shape charge and makeshift detonator Sam carried and took up positions in front of Sam. One of them stood by the west window, and the other two planted themselves in front of him and to the left.

The politicians sat glaring at him.

"Hello, fellas," Sam said, giving a little wave.

"You have some nerve, you little shit," Senator Lambert replied. "Do you realize the problems you caused us?"

"Well, yes, I do. I believe that's what led us all to this fateful day."

Lambert narrowed his eyes at Sam. "What do you expect to achieve?"

"Me? I've achieved everything except for one last item on my to-do list. My question is what do you plan on accomplishing by blowing up the Capitol building other than killing a bunch of politicians?" Sam looked around the room before continuing, "You've lost, gentlemen. Right now, a division of the military headed by General Forrest is securing all outlying areas of the city."

The group of men sat smirking at each other. Lambert spoke up.

"You think you have changed anything with this stunt today? You delayed the inevitable. You see those?" said the senator, pointing at the hounds. "They are the beginning of something much greater—a new future."

"If you mean these violations of the autonomous robot and artificial intelligence treaty, yes, I've become quite intimate with their function. My question is this: why blow up the building? Are you so stupid to think an escape route here exists I don't know about?"

"We have a means to escape. Don't you worry about that," The Speaker of the House replied.

"Are you talking about the tunnel leading from here to the Supreme Court building? Yeah, I stationed fifty soldiers at the other end of that."

Senator Patterson spoke up. "Bryan, you said nobody knew about the tunnel and we'd be able to evacuate during the distraction!"

"Shut up, you idiot. He's playing us," Lambert said. "Besides, I still have this." Senator Lambert patted the detonator on the table.

Sam stopped time, snatched the detonator the senator patted a moment before, and clicked his portion into the bottom of the device that had been sitting in front of the senator.

"You mean this detonator?"

Sam held out the shape charge and held it up in front of him. The senator glanced at the place where he'd patted the detonator, but it was gone.

"How in the hell did you get that?"

"I could go into a long story about my skills and where they come from, but I'd rather finish this last item on my list."

Sam lofted the shape charge into the air. Jake had already taken two steps towards Sam before he threw the explosive into the air. In a split-second decision, he changed direction and headed for a side door into another room as Sam pulled the trigger on the detonator.

The shape charge went off, blowing Sam through a door behind him and into another room. The blast in front of the charge obliterated the two hounds and the men in front of it.

The guard by the window was blown through the glass and fell three stories to the lawn below. The rest of them were crushed by the concussive force of the blast. Jake was thrown through the door to the room adjacent to where Sam had landed.

After the smoke and debris settled, Sam pushed himself up through a mass of broken plaster and wood, half blind and deaf from the blast. Blood trickled from his ears as he stumbled out of the door, looking at what was left of the senators. He rummaged around and found his sword in the debris by the door and slung it back on.

He turned and had just made it to the doorway when

Jake slammed into him from behind. They both stumbled through the balcony door to the Senate Chamber, down the stairs to the railing.

Jake pulled a sword and came down with a hard slash. Sam parried the blow with his right hand to Jake's forearm, and the sword sliced through the railing like it wasn't there. Sam choked from the debris dust, his ears rang, and his eyes were filled with tears from the blast. He locked on to Jake's arm and pulled them both over the railing.

They fell the twelve feet to the Senate floor below. Jake landed hard on the floor while Sam's fall was absorbed by a desk he landed on. It snapped under his weight, and he rolled onto his feet, still coughing and blinking.

He pulled his sword out while wiping the tears from his eyes. Sam forced his eyes open and pulled out the two contacts protecting his eyes from the sunlight. That made an instant difference. He was half deaf from the blast. His vision cleared in time to see Jake rushing at him with both swords.

Sam grabbed a hold of a nearby desk, lifting it in front of him as he blocked one sword with his own and the other cut through the desk, almost taking his fingers with it.

Jake held the swords up for Sam to look at.

"You like them? I had them made from the schematics I stole from you before I turned on Reboot."

Jake flourished the two gleaming blades to Sam as the two faced off.

"You realize you killed hundreds if not thousands of employees with that stunt, not to mention my foster parents?" Sam said.

"No! I am not responsible for that. Reboot killed those employees the moment it decided to stick its nose into politics. I brought about justice. I protected the Constitution of the United States from traitors!"

Sam shook his head sadly. "You've condemned yourself in the eyes of the Lord, Jake."

"Shut up with that garbage! There is no God that made man. That was dribble created as propaganda to cloud the minds of the citizens. Politicians and religion lead everyone around like sheep in the name of some fictitious God. You've pumped that crap into me ever since we were paired together. I tolerated it, nothing more. We were like brothers, Sam."

"Yeah, and you turned away from me because of jealousy and greed."

Jake jumped at Sam with two quick slashes that followed up into a low sweeping double slash at Sam's legs. Sam parried both high strikes and jumped over the low slash while he flicked his sword across Jake's cheek.

The blade was so sharp Jake didn't think he got nicked until he felt the blood trickling down his neck.

He growled in rage and lunged at Sam with well-practiced feints and slashes. Sam was still recovering from the blast and remained on the defensive.

Jake feinted at Sam.

Sam slid up inside Jake's defense and put a hard elbow into his ribs. Jake grunted in pain from the blow. In return, Jake was able to get an edge across Sam's shoulder, flaying the flesh open to the bone.

Sam winched from the pain, but he kept on the pressure and immobilized Jake's arm until Jake had no choice but to drop the sword in his left hand. Jake swung the sword in his right hand toward Sam's legs. Sam blocked the attack with his own sword.

Jake changed his tactics in return. He dropped his

second sword and judo-flipped Sam onto his back. Jake locked Sam's wrist, disarming him. He tried breaking Sam's wrist, but Sam's dense bones prevented him from getting enough leverage to accomplish that before Sam leg-swept him.

They both scrambled to their feet. Blood was running down Sam's arm and dripping off his hand. The damage to his shoulder was significant, and he had difficulty holding his arm up. Jake sensed his advantage and rushed at Sam, pinning him up against the marble desk at the head of the room.

Jake pulled a knife from his belt and thrust it at Sam. The knife plunged through Sam's palm, but Sam managed to latch on to Jake's hand, effectively disabling the knife.

Jake dug his thumb and fingers into the gash on Sam's shoulder with his free hand. Sam let out a howling scream. He went to the only place he could with that rage, triggering his second strength feat in the same day.

His sudden surge of strength shocked Jake. In less than two seconds, Sam crushed Jake's hand where the knife was impaled through in his vise-like grip.

Sam pivoted during the move and brought his hand from his injured arm up under Jake's arm and up to behind his head, preventing him from blocking what was coming next. Sam screamed in pain as he thrust Jake's head and face into the marble desktop, crushing both. Sam blacked out a second later, slumping over face-first to the floor of the Senate Chamber.

CHAPTER FORTY-FIVE

S am opened his eyes. He found himself sitting at the bench in the grove where he'd met Jesus for the first time. Jesus sat across from him, smiling.

"Am I dead?"

"No, my son, you are not."

"Did we win?"

Jesus chuckled. "This battle."

Sam relaxed. "I hoped you would say that. Wait ... *this* battle?"

"As I indicated in our previous meeting, you are being prepared for something much bigger and more difficult. You have progressed in your skills as expected. The difficulties that lie ahead will require you to continue to improve and master those skills."

"How much time do I have? Where is the threat coming from? Why are you looking at me like that?"

Jesus reached forward and rested his hand on Sam's shoulder. A calm came over him like a warm blanket.

"Don't fret. Enjoy this victory. I have faith in you the same as you have faith in Me."

"I don't know if I would consider those the same. I mean, you're Jesus Christ-our savior."

"So, you think your faith in Me is different from faith in anything else? What of Laura Cinder? You have faith in her to do what is right and just, faith that she will not lead you astray or come to harm you. How about Alyssa? Your foster parents? The faith you have in each other is no different from faith in Me.

"Humanity creates bonds of faith from those who are important in their life. The majority cannot see the love I have for them. Therefore, their faith is not physical as it is with those they can see, hear, and touch. Your faith in Me was no different before you met me. Now that you can put a face and a physical form with that faith, it is much stronger than any faith you experienced in the past. However, you are an exception.

"Remember this, Samuel, the rest of the world has been through a lot. But they do not have the luxury of knowing what you do, and nothing you say to them will change that."

Sam realized every word Jesus said rang true in his heart. He was special. It wasn't just the blessings he received from Jesus that made him so. He knew without a doubt that Jesus, God, and the heavens existed. He might be the one person on Earth who glimpsed the reality of the world and the heavens above. For everyone else, it fell to faith to keep them going.

"I see you understand now."

Sam nodded. "What happens now?"

"Now you will wake up in the hospital. You will be waking from a coma the doctors cannot explain."

"Will I have a splitting headache?"

Jesus smiled. "Not this time."

The light coming from Jesus intensified until everything was awash in it.

"Good luck, my son," Jesus said as he faded from view.

When colors started coming back, Sam could make out the walls of a hospital room. The news displayed on the wall in front of him, but he couldn't hear anything.

In his peripheral vision, he picked out Alyssa napping in a chair. His hearing faded in with the newscast and the beeping of the machines that monitored him.

It was at that moment he realized Laura Cinder stood behind the person who had just walked up to the podium. They were standing on the steps of the Capitol building. The person at the podium cleared his throat. Sam recognized him from his graduation.

"My name is Robert Bartram. I am a historian and public relations specialist for Reboot. As many of the citizens watching today have come to realize, our government is no more, at least the government that existed at the beginning of the day. The corrupted, greedy, power-hungry, and broken form of what the creators of the Constitution of the United States always feared it might become is gone

"The sad part of that statement is this city was built on having proper representation. Every person here was hired by the people to represent them. In the years leading up to the last world war, our democracy ceased to exist. On the surface it read as a democracy, but below the surface, the government remained nothing of the sort.

"To achieve anything in the capital of our nation required great sums of money. I am standing here today as evidence of that fact. Reboot facilitated the removal of the government by using the need for power and money against it.

"Many of you are likely thinking of what is to happen

next. Put simply, power will be returned to the people. Our country and other countries of the world are shells of their former selves, but that doesn't mean we need to live from one day to the next just getting by.

"The money wasted since Reboot developed a cure for the pandemics ravaging our world could have rebuilt every city in our country and repaired every piece of infrastructure we took for granted before the war broke out. That never happened. Instead, nature has reoccupied a vast majority of our cities, roads, and bridges. Travel between cities is no longer something that can be done over a week-end. If you don't live in a city, then you are on your own. That is how the government preferred it.

"The government we plan on implementing is the same one the creators of the Constitution tried in vain to put into place. We are going to give democracy another try. Here are a few of the new regulations being enacted before government once again is given the green light.

"First, the Constitution is going to be wiped back down to the original listing and then updated to this age. Second, term limits are no longer going to be exclusive teo the executive branch of government. All of congress and the Supreme Court justices will have term limits put into place. Third, the Democratic and Republican parties are to be disbanded and stripped of their power. Specific guidelines to party creation and what a party can and cannot do will be established prior to any new party affiliation being determined.

"Fourth, no future election campaigns will be funded outside of the government. A set amount will be established, and certain privileges will be given to those who are campaigning to prevent illegal donations. Fifth, no donations of any kind from lobbyists or foreign governments will be accepted by any party or government official."

Sam sat listening to the stream and smiled. They accomplished the impossible. He glanced over to Alyssa, who still snoozed in her chair.

"What does a guy gotta do to get a hug around here?" he said, his voice hoarse.

Alyssa jumped out of her chair as if she sat on a cattle prod.

"You're awake!" she gasped.

She leaped over to the bed and hugged him.

"So, how did I get here?"

She stood up and looked serious for a moment.

"When the explosion ripped through the west side of the Senate wing, the military and I went to investigate. I found you face-down on the senate floor next to Leo's—er, I mean Jake's body. At first, I thought you were dead too, but you had a pulse, so I called for an emergency evac to rescue you."

"How long was I out?"

She glanced at her wrist.

"Four days, seven hours, and twenty-six minutes. Not that I'm counting."

Sam swung his legs out of the bed and stood. He grabbed his shoulder. The wound was healed and no scar was visible.

About that time, Laura came walking in on her phone with some assistant tagging along behind her like a puppy. She paused when she realized Sam was awake.

"One moment," she said into her phone, then spoke to Sam.

"Look at you! Awake and standing!"

She half-skipped, half-jumped over to embrace him.

"Thank you, Sam. For everything. Thank you, thank

you, thank you." She kept repeating into his ear as he gave her the biggest bear hug.

Sam realized at that moment the two people he had the most faith in stood in the room with him.

A storm might be looming, but at least this moment could be relished. He pulled Alyssa into the hug. Life didn't get better than this.

EPILOGUE

Robert sat back in his chair, a bit out of breath after recounting all this information. He closed his eyes for a moment and listened to the hum of the machine that was keeping him alive.

Drew looked over at him with a worried expression.

"Would you like to come back next week and continue your story?"

Robert opened his eyes and looked back at him.

He sighed. "I don't have a week."

Robert grunted as he adjusted himself in his seat. The biographer blushed a bit out of embarrassment as he took in what Robert said.

"I've lived a very long life, Drew. Do not feel sorry for me. I will continue where I left off, but that will need to wait until tomorrow. For now, I need to rest."

"I understand. When can I expect you tomorrow?"

"I think it's best that you make the trip to see me tomor-row. I have to rest for more hours than I'm awake. The doctors will contact you when I'm going to be available."

Drew smiled at Robert.

"We will be ready for you, sir. Thank you again for doing this."

Robert smiled a halfhearted smile as the nurse gathered his things and pushed him back out to the waiting medical transit that would take him back to the hospital.

Drew finalized a few things on his tablet to archive the data away for later review.

He grabbed a cup of coffee and brought it to his lips, taking a sip. It had gone cold.

Drew sat it back down and stretched. He felt wiped after that session. How Robert had spoken for that long was amazing, considering his condition.

Sam was a very interesting person, and the things he had experienced up to this point astounded Drew. Yet, this was just the beginning. His conversation with Jesus and being the tip of the spear accounted for more than just overthrowing the U.S. government.

He felt there was a lot more to the South African story coming in the next session.

Everyone knew of South Africa. He didn't think anyone knew of Sam's involvement with them. He was excited to hear details about that.

He yawned and stretched again, scrubbing his hands over his face. He turned off his tablet and the rest of the gear in the room.

He prayed Robert had a good rest and was able to schedule something with him sooner than later. He needed to prepare for a short notice follow-up interview. As tired as he was, he decided he would get the recording gear ready in case the call came at an odd hour.

Sam was a shooting star and seemed to be burning

bright. Drew was excited to find out what happened to his childhood hero. The journey was turning out to be even more fulfilling than he'd hoped.

ACKNOWLEDGMENTS

It may be a bit cliché, but I'd like to thank my mother. Her drive and spirit in the face of any obstacle continuously inspires me and others to achieve more than they would without her example. I can only aspire to be more like her, giving and selfless.

A big thank you to my editor Casey at Thoth Editing. My book would not be what it is without your guiding ways.

Another thank you goes out to my book cover designer Nelly at Pix Bee Design for collaborating with me to create a cover that I think best represents the contents of the book.

To the reader, I hope you enjoy reading my story as much as I did in creating it. My goal is to entertain and inspire, and this is my attempt at both.